Domestic Politics and International Relations in US–Japan Trade Policymaking

The GATT Uruguay Round Agriculture Negotiations

Christopher C. Meyerson

palgrave
macmillan

First published 2003 by
PALGRAVE MACMILLAN
Houndmills, Basingstoke, Hampshire RG21 6XS and
175 Fifth Avenue, New York, N.Y. 10010
Companies and representatives throughout the world

PALGRAVE MACMILLAN is the global academic imprint of the Palgrave Macmillan division of St. Martin's Press, LLC and of Palgrave Macmillan Ltd. Macmillan® is a registered trademark in the United States, United Kingdom and other countries. Palgrave is a registered trademark in the European Union and other countries.

ISBN 1–4039–0799–4

This book is printed on paper suitable for recycling and made from fully managed and sustained forest sources.

A catalogue record for this book is available from the British Library.

Library of Congress Cataloging-in-Publication Data
Meyerson, Christopher Cortlandt.
Domestic politics and international relations in US–Japan trade policymaking: the GATT Uruguay Round agriculture negotiations / Christopher C. Meyerson.
    p. cm. – (International political economy series)
  Includes bibliographical references and index.
  ISBN 1-4039-0799-4
    1. United States – Commercial policy – Decision making. 2. Japan – Commercial policy – Decision making. 3. Produce trade – Government policy – United States. 4. Produce trade – Government policy – Japan. 5. Uruguay Round (1987–1994) 6. General Agreement on Tariffs and Trade (Organization) 7. United States – Foreign economic relations – Japan. 8. Japan – Foreign economic relations – United States. 9. United States – Politics and government – 20th century. 10. Japan – Politics and government – 20th century. I. Title. II. International political economy series (Palgrave Macmillan (Firm))
HF1455.M49 2003
382'.41'0952–dc21   2003042964

10 9 8 7 6 5 4 3 2 1
12 11 10 09 08 07 06 05 04 03

Printed and bound in Great Britain by
Antony Rowe Ltd, Chippenham and Eastbourne

Contents

Preface and Acknowledgements

Linked to efforts to promote trade liberalization through international trade negotiations is the need to better understand the relationship between domestic politics and international relations in trade policymaking.

In order to address this need, this book develops a contextual two-level game approach to analyzing trade policymaking that strives to improve upon Robert Putnam's 1988 two-level game approach and to define more systematically, with reference to the existing literature, the nature of and relationship between domestic politics and international relations in one particular area – trade policymaking.

Chapter 1 develops the contextual two-level game approach to analyzing trade policymaking, setting out a series of 'if, then' testable, falsifiable hypotheses that address the question: To what extent does domestic politics affect the agreement reached in an international trade negotiation?

Chapter 2 demonstrates how the approach to analyzing trade policymaking and the hypotheses set out in Chapter 1 can be used to explore the relationship between domestic politics and international relations in American trade policymaking, Japanese trade policymaking, and US–Japan trade negotiations. Chapter 2 concludes by discussing how the contextual two-level game approach can be further developed by examining a particular case study – American trade policymaking, Japanese trade policymaking, and US–Japan trade negotiations during the GATT Uruguay Round agriculture negotiations (1986–94).

Chapters 3, 4, and 5 test variations of the hypotheses set out in Chapter 1 in order to analyze the relationship between domestic politics and international relations in American trade policymaking, Japanese trade policymaking, and US–Japan trade negotiations during the GATT Uruguay Round agriculture negotiations.

Chapter 6 concludes by assessing the effectiveness of the contextual two-level game approach, and by suggesting areas for further research.

The researching and writing of this book involved extensive study in numerous libraries in the United States and Japan, and the exchange of ideas with various scholars, government officials and other individuals knowledgeable concerning the particular theories and negotiations discussed in this book. Research is included based on derestricted GATT documents related to the negotiations that have recently become available to the public.[1]

This book contributes to the findings of previous studies that concern the GATT Uruguay Round agriculture negotiations, such as William Avery, 1993a; John Breen, 1993, 1999; K. A. Ingersent, A. J. Rayner and R. C. Hine, 1994; and Remco Vahl, 1997. The book also sets out interpretations which are different from those of Robert Paarlberg 1993, 1997; David Rapkin and Aurelia George 1993; and Yoichiro Sato 1996, who have used Putnam's two-level game approach to examine the relationship between domestic politics and international relations in American trade policymaking, Japanese trade policymaking, and/or US–Japan trade negotiations during the GATT Uruguay Round agriculture negotiations.

I am very grateful to the many individuals who helped me during the researching and writing of this book, although it is impossible to list all of the individuals who assisted me with this work over the last few years. I am particularly grateful to those who became interested in turning this project into a book, especially Timothy Shaw, general editor of the International Political Economy series in which this book appears, and Amanda Watkins, my commissioning editor at Palgrave. I would also like to thank Keith Povey for his copy-editing.

I began working on the theoretical conceptualization and overall structure of this book at Columbia University during a seminar taught by Professor Douglas Chalmers. During 1991, I decided to focus on the GATT Uruguay Round agriculture negotiations. Professors Gerald Curtis and Richard Nelson guided me through the beginning of this project, and Professor Mark Kesselman and the Dissertation Review Committee of the Department of Political Science of Columbia University provided useful comments. Because I received a Japanese Government Ministry of Education (*Monbushō*) Scholarship to study at Kyoto University Faculty of Law, I began the research and writing of this book by focusing on the role of Japan in the ongoing GATT Uruguay Round agriculture negotiations. In March 1992, I was admitted to a two-year LLM program under the supervision of Michio Muramatsu, Professor of Public Administration at Kyoto University Faculty of Law. As part of my LLM requirements, I completed an LLM thesis in Japanese concerning Japan's role in the Uruguay Round agriculture negotiations (Christopher Meyerson, 1994). I am grateful for the comments of Professors Michio Muramatsu, Masataka Kōsaka and Kengo Akizuki during the defense of my LLM thesis, as well as to the students of Professor Muramatsu's Public Administration Seminar for their comments concerning my research. After defending my Kyoto University LLM thesis successfully in March 1994, I was admitted to the Doctoral Program at Kyoto University Graduate School Faculty of Law, and my Japanese Government scholarship was extended to March 1997 to cover the three-year residency requirement of the Doctoral Program of Kyoto University Graduate School Faculty of Law.

While in Japan, I increased my understanding of US–Japan relations by teaching seminars entitled 'Legal Issues of US–Japan Relations' and 'The Political Economy of the United States and Japan' during the Spring and Fall 1996 semesters to students from American colleges enrolled in the Associated Kyoto Program at Doshisha University, which is an academic-year abroad program for students from a consortium of American colleges, which included at that time Amherst College, Williams College, Middlebury College, Pomona College, Oberlin College, Wesleyan University and other colleges and universities.

In April 1997, I returned to Washington, DC, where I continued to work on this book, and am particularly grateful to Professor Helen Milner, as well as Professors Jack Snyder, Gerald Curtis, Richard Nelson, Andrew Nathan and Charles Cameron of the Political Science Department of Columbia University for guiding me through revisions.

I am also grateful to those individuals who provided comments concerning summaries of this book or sections of this book that I presented at various conferences and meetings, including David Arase, Theodore Cohn, Donald Dalton, Mine Eder, H. Richard Friman, Fen Hampson, A. Christian Harth, Masaru Kohno, James C. Roberts, Leonard Schoppa, Robert Uriu and Duncan Wood. A related paper concerning the George H. W. Bush Administration's role in the Uruguay Round negotiations was presented at the April 1997 Hofstra University Conference on the Presidency of George H. W. Bush (Meyerson, 2002). I am grateful for the comments of Ambassador Carla Hills concerning that paper.

I would like to thank John Dickson of the World Trade Organization for providing me guidance concerning the appropriate citation of GATT Uruguay Round documents. Some sections of this book appeared previously in Meyerson, 2000, for which I was awarded the International Studies Association International Political Economy Section 2000 Junior Scholar Award, and I am grateful for Sean S. Costigan, Senior Editor of Columbia International Affairs Online for granting me permission to reprint sections of that paper. Related conference papers, some of which are listed in the Bibliography, include papers that I presented at the International Studies Association 1999 and 1997 annual meetings (Meyerson, 1999, 1997b), the American Political Science Association 1997 and 1996 annual meetings (Meyerson, 1997d and 1996c), the Association for Asian Studies 1997 and 1996 annual meetings (Meyerson, 1997a and 1996a), the Association of Japanese Business Studies 1997 and 1996 annual meetings (Meyerson, 1997c and 1996b), and the 1996 International Studies Association-Japan Association of International Relations Joint Convention (Meyerson, 1996d).

I would also like to thank my family for their patience and understanding during this project.

This is an ambitious project, and I would be very interested in receiving any advice and suggestions that readers have concerning this book.

CHRISTOPHER C. MEYERSON

List of Abbreviations

Add.	Addendum
ASEAN	Association of South East Asian Nations
Cong.	United States Congress
Doc.	Document
EC	European Community
edn	edition
FA	Final Act
G-7	Group of Seven
G-8	Group of Eight
G-14	Group of Fourteen
GATT	General Agreement on Tariffs and Trade
GDP	Gross Domestic Product
GNG	Group of Negotiations on Goods
H.	US Congress, House of Representatives
JSP	Japan Socialist Party
LDP	Liberal Democratic Party
LLM	Master of Law
MA	Market Access
MIN	Ministerial
MITI	Ministry of International Trade and Industry
MOSS	Market-Oriented Sector-Selective
MTN	Multilateral Trade Negotiations
NAFTA	North American Free Trade Agreement
NFU	National Farmers Union
NG5	Negotiating Group on Agriculture
No.	Number
NUR	*News of the Uruguay Round of Multilateral Trade Negotiations*
OECD	Organisation for Economic Co-operation and Development
Rept.	Report
Rev.	Revision
S	US Congress, Senate
sec.	section
sess.	Session
TNC	Trade Negotiations Committee
TRIPS	Trade-Related Aspects of Intellectual Property Rights
US	United States

USA   United States of America
USDA   United States Department of Agriculture
USITC   United States International Trade Commission
USTR   United States Trade Representative
vol.   volume
W   Working Draft
WTO   World Trade Organization

1
Introduction

The growing number of countries that have become members of the World Trade Organization (WTO), and the expanding number of trade-related issues that are of concern to citizens of various countries, have led to an increased need to better understand the relationship between domestic politics and international relations in trade policymaking.

While not many authors have attempted to explore the relationship between domestic politics and international relations in trade policymaking, several notable efforts have been made in recent years to explain in general the relationship between domestic politics and international relations. Among these efforts,[1] this book focuses on the two-level game approach originally set out in Robert Putnam's 1988 article, 'Diplomacy and Domestic Politics: The Logic of Two-Level Games' in *International Organization*.

Putnam's two-level game approach combines domestic-oriented and international-oriented theories with theories of the negotiating process. Putnam conceptualizes the negotiating process between two countries as a two-level game structure that occurs in two simplified stages: '1. bargaining between the negotiators, leading to a tentative agreement; call that Level i [and] 2. separate discussions within each group of constituents about whether to ratify the agreement; call that Level ii' (Putnam, 1988, p. 436).

As James Caporaso notes, Putnam's two-level game approach is 'uniquely appropriate for situations in which there are at least two distinct governments at odds over some issue, but where negotiations could in principle yield joint gains'. Putnam's two-level game approach is most applicable, not when the international system or domestic politics is paramount, but when the chief of government 'simultaneously represents domestic constituents and searches for acceptable deals with counterparts abroad' (Caporaso, 1997, pp. 585-6).

In Putnam's conceptualization, at the domestic level, 'domestic groups pursue their interests by pressuring the government to adopt favorable policies, and politicians seek power by constructing coalitions among those groups'. At the international level, 'national governments seek to maximize their own ability to satisfy domestic pressures, while minimizing the adverse consequences of foreign developments' (Putnam, 1988, p. 434).

Putnam defines the '"win-set" for a given Level ii [domestic] constituency as the set of all possible Level i [international] agreements that would "win" – that is, gain the necessary majority among the constituents – when simply voted up or down'. The size of the win-set is affected by 'Level ii [domestic] preferences and coalitions; Level II [domestic] institutions; [and] Level I negotiators' strategies'. In this model of two-level negotiations, 'the chief negotiator is the only formal link between Level i and Level ii' (Putnam, 1988, pp. 437, 441–2, 456–7).

While Putnam's two-level game approach represents a notable advance in conceptualizing the relationship between domestic politics and international relations, Putnam does not link his two-level game approach very effectively to related domestic- and international-oriented theories.

At the domestic level, while Putnam mentions that domestic preferences and coalitions and institutions are involved in policymaking, he does not elucidate how to determine which constituents, institutions, preferences and coalitions are important. Nor does he explain how to assess the relationship between them. Furthermore, the international level of Putnam's two-level game approach consists of the simplified conceptualization of 'bargaining between the negotiators, leading to a tentative agreement' (Putnam, 1988, p. 436), which neglects considerations of the influence of international regimes, interdependence, cooperation, countries' positions in the international system and other factors that are emphasized in the literature of international relations.

Putnam concludes his 'Diplomacy and Domestic Politics' article with a checklist of 'several significant features of the links between diplomacy and domestic politics'. Among the features that Putnam notes are 'the possibility of synergistic issue linkage, in which strategic moves at one game-table facilitate unexpected coalitions at the second table'; 'the paradoxical fact that institutional arrangements which strengthen decision-makers at home may weaken their international bargaining position, and vice versa'; 'the importance of targeting international threats, offers, and side-payments with an eye towards their domestic incidence at home and abroad'; and 'the potential reverberation of international pressures within the domestic arena' (Putnam, 1988, p. 460). While this checklist sets out an interesting research agenda, it does not lend itself easily to systematic application to other cases. Thus some have commented that Putnam's two-level game approach does 'not constitute a theory with testable hypotheses' (Milner, 1997, p. 4, citing Peter Evans *et al.*, 1993).

While many studies – for example, Avery 1993a (especially pp. 4–16) – claim to use some form of Putnam's two-level game approach to analyze an event that involves both domestic and international factors, there have been only a few systematic efforts to improve upon some of the weaknesses of Putnam's two-level game approach.

Five years after 'Diplomacy and Domestic Politics' was published, Putnam and a number of other scholars published a volume entitled *Double-Edged Diplomacy: International Bargaining and Domestic Politics. Double-Edged Diplomacy* (Evans *et al.*, 1993) tries to develop, enhance and expand Putnam's two-level game approach. In the introductory chapter, Andrew Moravcsik notes that three theoretical building blocks are needed in order to further develop Putnam's two-level game approach: 'specifications of domestic politics (the nature of the "win-sets"), of the international negotiating environment (the determinants of interstate bargaining outcomes), and of the statesman's preferences' (Moravcsik, 1993, p. 23). In the last chapter of *Double-Edged Diplomacy*, Evans makes a couple of observations concerning the relationship between domestic politics and international relations, noting that when 'the costs are concentrated and the benefits diffuse, agreements are usually, though not invariably doomed', that there is 'no relationship between the extent of enfranchisement and the propensity to conclude agreements'; and that there 'is no relationship between the presence of transnational actors in a particular arena and the likelihood that an attempted agreement will succeed' (Evans, 1993, pp. 399–400).

Schoppa's 1997 *Bargaining with Japan* attempts to complete what Putnam began, by connecting systematically various negotiation strategies to a revised version of Putnam's theory of domestic politics, 'in order to generate hypotheses about when these strategies are likely to prove effective'. Focusing on the US and Japan, Schoppa examines how institutional structures have influenced 'the behavior of various actors within their domestic polities', and how 'media and groups of experts in specific areas of policy ("policy communities") play very important roles' (Schoppa, 1997, pp. 30-2). Developing theories from Putnam's checklist, Schoppa sets out six hypotheses concerning the relationship between domestic politics and international relations. These hypotheses concern reverberation, synergistic threats, synergistic linkage, participation expansion, alternative specification, and tying hands (Schoppa, 1997, pp. 32–48). Using this theoretical structure, Schoppa attempts to determine what American pressure on Japan can and cannot do, examining US–Japan trade negotiations during the Structural Impediments Initiative and the Clinton Administration Framework Talks. (For a critique of Schoppa, see Norio Naka, 1996, pp. 4–7.)

Milner's 1997 *Interests, Institutions, and Information* uses rational choice theory to improve the theoretical conceptualization of Putnam's two-level game approach and to develop testable hypotheses. Milner develops an explicit theory to answer the question of why nations cooperate with each other by examining when and why advanced Western industrial countries were unable to cooperate in the post-Second World War period in trade, monetary and fiscal policy (Milner, 1997, pp. 5–6). In the model that Milner develops, the domestic level includes three actors – the executive, the legislature, and domestic interest groups – who share power over decisionmaking. The international level is conceived of 'as an anarchic environment, free of institutions, involving competition between two unitary states'. Milner argues that domestic politics makes cooperation more difficult for countries. Milner states that 'Three internal factors condition a state's ability to cooperate: *the structure of domestic preferences, the nature of domestic political institutions, and the distribution of information internally.* These factors determine which domestic players are involved in policy making, what powers each has in this process, and how these actors' policy preferences differ'. Using this model, Milner derives hypotheses that link domestic politics to the negotiation and ratification of international agreements. Milner argues that 'cooperation among nations is less plagued by fears of other countries' relative gains or likelihood of cheating than it is by the *domestic distributional consequences* of cooperative endeavors' (Milner, 1997, pp. 70–2, 233–4).

The contextual two-level game approach to analyzing trade policymaking

This book, like Schoppa's 1997 *Bargaining with Japan* and Milner's 1997 *Interests, Institutions, and Information*, strives to improve on Putnam's two-level game approach by developing and testing a series of testable hypotheses concerning the relationship between domestic politics and international relations. Like Schoppa's *Bargaining with Japan*, this book attempts to refine the two-level game approach by focusing on trade policymaking and US–Japan trade negotiations. But, in contrast to Schoppa's *Bargaining with Japan* and Milner's *Interests, Institutions, and Information*, this book starts from a broader and more comparative perspective, and places more emphasis on clearly identifying and defining, with reference to the existing literature, the key domestic and international variables involved in one particular issue area – trade policymaking. Particular attention is paid to Ellis Krauss's call in his article in Evans *et al.*, 1993 *Double-Edged Diplomacy* for re-evaluating the role of domestic actors in order to improve the understanding of the relationship between domestic politics and international relations (Krauss, 1993, p. 293).

In the paragraphs that follow, the key domestic and international variables involved in trade policymaking are set out, while striking a balance between theoretical parsimony and the complexity of the trade policymaking process. The variables' relationships to each other are discussed, and hypotheses are developed. In Chapter 2, the contextual two-level game approach is used to explore differences and similarities in the trade policymaking of certain countries – namely, the United States and Japan. In Chapters 3 to 6, the contextual two-level game approach to analyzing trade policymaking is developed by examining a particular trade negotiation. Focusing on both variables and case studies to develop such an analytic approach avoids making simplifying causal assumptions that can result from just focusing on variables. Such a synthetic strategy is also 'holistic – so that the cases themselves are not lost in the research process – and analytic – so that more than a few cases can be comprehended and modest generalization is possible' (Charles Ragin, 1987, pp. xiv, 51–2, 67–71, 82–4).

Since an extensive literature that focuses on the relationship between domestic politics and international relations in trade policymaking does not exist, it is useful to begin constructing such an approach by reviewing briefly the existing domestic- and international-oriented approaches to analyzing foreign economic and trade policymaking. Foreign economic and trade policymaking involves not only trade-related issues, such as negotiations concerning tariffs and trade barriers, but also finance-related issues, such as foreign direct investment and foreign exchange, as well as foreign aid and the economics of defense-related issues.

The domestic level

Existing domestic-oriented approaches to explaining foreign economic and trade policymaking include state-centered, society-centered, ideas-oriented and rational choice approaches, as well as approaches that focus on the relationship between the executive, the legislature and interest groups. Most such approaches provide some way to analyze changes in foreign economic policymaking that occur over time, as discussed in G. John Ikenberry (1988, pp. 229–36).

A state-centered approach,[2] which is sometimes called an institutional approach or domestic structure approach, can be used to argue that trade policy results from government officials actively pursuing autonomous goals, or to argue that trade policy is shaped and constrained by the organizational structure of the state.

A society-centered approach can be used to argue that trade policy results from 'the demands placed on government by private groups, sectors, or classes within the national political system' (Ikenberry, 1988, p. 219). In the society-centered approach, the analysis of the role of interest groups is particularly important, and government institutions provide an arena in which groups compete and government institutions exert less significant influence on the decisions that are made (Ikenberry *et al.*, 1988b, pp. 7–9). Related to the society-centered approach is analysis that focuses on the role of coalitions, such as Peter Gourevitch (1978b, p. 905) and Gourevitch (1986).

An ideas-oriented approach can be used to argue that trade policy is influenced by ideas such as protectionism and liberalism, as Judith Goldstein (1993) argues.

A rational choice approach can be used to argue that trade policy results from voters' preferences – for example, from political actors seeking to develop a trade policy which 'perfectly balances the many preferences of their constituents so that their chances of reelection are maximized'.[3]

The domestic level of the contextual two-level game approach to analyzing trade policymaking developed in this book consists of three variables – the state (bureaucracy),[4] the legislature, and other actors in the policymaking process – and the relationships between them. These variables are derived from the neo-pluralist model of democracy set out in David Held (1996, pp. 214–18). Held's neo-pluralist model is based largely upon the literature on neo-pluralism in the United States.[5] In the contextual two-level game approach, a state (bureaucracy), a legislature, and other actors in the trade policymaking process are assumed to exist to some extent in all countries in the world.[6] However, not all of the governments in these countries are assumed to be neo-pluralist. Rather, the assumption is made that in many countries the relationships between these three variables may resemble the relationships set out in the neo-pluralist model, in which the 'poor resource base of many groups prevents full political participation' and the 'distribution of socioeconomic power provides opportunities for and limits to political options.' As a result, 'unequal involvement in politics' creates an 'insufficiently open government' (Held, 1996, pp. 217–18).

Elements of the state (bureaucracy) are involved in the administration of imports and exports, as well as in trade negotiations with other countries. Politicians and administrators within the state (bureaucracy) often act as independent participants in the trade policymaking process. The major trade negotiator of a country in a particular trade negotiation is somewhere within the state (bureaucracy), but is not necessarily the chief of government.

The legislature's role in trade policymaking is often determined by constitutional rules that affect the set of laws and institutional arrangements governing trade policymaking. In many countries there may be formal or informal groups within the legislature that concern themselves especially with trade policymaking. The legislature's role in trade policymaking varies enormously from country to country and over time, and can involve, as Milner suggests, agenda setting, amendment power, ratification or veto power, the proposal of referendums, or side payments (see Milner, 1997, pp. 102–12).

Various other actors in the trade policymaking process, including interest groups, reflect the preferences of particular social forces, political groups or classes in society who struggle for influence in the trade policymaking process. Such actors can include importing and exporting companies, labor unions and other groups affected directly by trade policymaking, as well as the media, who can be some of the most important influences on trade policy. Producers facing competition from imports, and others interested in protectionism, may be more organized than exporters and consumers who favor free trade, as discussed in Lisa Martin and Beth Simmons (1998, p. 748).

Many diverse elements of the state (bureaucracy), the legislature, and other actors in the trade policymaking process, whose alliances may shift, participate in the trade policymaking process. The resulting trade policy emerges in the context of shifting coalitions that form between elements of the state (bureaucracy), the legislature, and other actors in the trade policymaking process. The formation of these coalitions is affected, as Milner suggests, by a number of factors including the 'policy *preferences* of domestic actors, the *institutions* for power sharing among them, and the distribution of *information* among them' (Milner, 1997, p. 11). Relations between the state (bureaucracy), the legislature, and other actors in the trade policymaking process vary among countries and across issue areas.

The international level

The international level of the contextual two-level game approach to analyzing trade policymaking includes a number of variables that can affect a country's trade policymaking. These are outlined in the paragraphs that follow, and include the international trade regime, the position of a country in the international political economy, interdependence, the evolution of capitalism, the period during which a country industrializes, changes in the trade policymaking of countries and within international trade institutions, a country's desire to cooperate, the number of countries involved in a negotiation, and negotiating strategies.

Through a series of international trade regimes, countries have mutually agreed to abide by and enforce certain 'principles, norms, rules, and decisionmaking procedures' (Stephen Krasner, 1982b, p. 1) around which their expectations have converged (Robert Keohane, 1984, pp. 49–132; Milner, 1992, p. 475) concerning their trade relations. As a result, international anarchy has been mitigated. The current international trade regime, which was established through the GATT and which continues in the recently created WTO, includes a series of norms concerning liberalization, reciprocity, nondiscrimination, the right to take certain measures, and economic development, as Jock Finlayson and Mark Zacher (1982, pp. 278–96) note. John Ruggie has called the international trade regime that was established in the wake of the Second World War through the General Agreement on Tariffs and Trade (GATT) in 1947 a 'compromise of embedded liberalism', a form of multilateralism 'predicated upon domestic interventionism' which was designed to be 'compatible with the requirements of domestic stability' (John Ruggie, 1982, pp. 393, 399).

Mutual tariff reduction became institutionalized in the GATT through a series of negotiating rounds that involved a growing number of countries, as Bernard Hoekman and Michel Kostecki (1995, pp. 12–20) describe. During the various postwar GATT negotiating rounds, GATT member countries exchanged offer and request lists, and reached a series of agreements between themselves to mutually reduce tariffs and other trade barriers. These agreements were combined into a single agreement at the conclusion of each negotiating round, as Robert Baldwin (1988b, p. 192) notes. The last GATT negotiating round was the Uruguay Round (the subject of Chapters 3 to 6 of this book), which has been followed by the WTO Doha Agenda negotiations. Various issues have been incorporated into the GATT–WTO trade regime through the processes of agenda formation, negotiation, and operationalization discussed in Oran Young (1998, pp. 4–27, 188–94) see also Oran Young (1994). As a result of the treaty-like nature of the GATT, GATT decisionmaking processes were much less public than decisionmaking processes in the United Nations and other international organizations (Gerard Curzon and Victoria Curzon, 1973, pp. 300–2). As different issue areas, such as agriculture, have come to be incorporated into the GATT–WTO trade regime, specific regimes, such as a global agricultural trade regime, have come to be vested within the international trade regime (Cohn, 1993a, p. 37). Regional trade arrangements, such as the European Union and the North American Free Trade Agreement (NAFTA) have also emerged.

The relative institutionalization of the GATT and the WTO, as discussed in John Jackson (1988, pp. 15–20), as well as the extent to which a network of knowledge-based experts develops concerning a particular trade issue (William Drake and Kalypso Nicolaidis, 1992; Peter Haas, 1992, pp. 2–3) can affect a country's trade policy. Domestic actors who cooperate with or oppose countries' representatives at the GATT (or now the WTO) have sometimes been able to use the weak institutionalization of the GATT and the WTO to their advantage. At other times, the secrecy surrounding the GATT and the WTO has inhibited greater political participation and led to protests by concerned interest groups.

A country's position in the international economy can affect its trade policy (Friman, 1990, pp. 11–15). Some assert that dominant or hegemonic states (such as the United States during the postwar period) have a strong preference for liberal trade regimes and 'possess the power to create and maintain such regimes, either by providing collective goods or by coercing reluctant states to participate' in such regimes.[7] The size of a state in relation to the world economy can also affect a nation's trade policy (Peter Katzenstein, 1985).

Transnational relations, modernization and interdependence can also affect a government's freedom of action in designing and implementing its trade policy.[8]

International capitalism, as dependency, core-periphery and imperialism theories[9] suggest, and market conditions – as John Odell (2000, pp. 47–69) notes – can constrain, or even help to determine, the trade policy options available to a country. A country's trade policy may be affected strongly by the character of the world economy when a country attempted or attempts industrialization.[10]

Diffusion theories imply that new ways of handling trade policymaking that are adopted in one country or at the GATT or WTO can influence the development of a country's trade policy, as Gourevitch (1978b, pp. 891–2) notes.

The extent to which one country cooperates with another in trade can be affected by each country's desire to realize absolute or relative gains, to continue to interact indefinitely with the other country, and by the perception of power imbalances between the countries (Milner, 1992, pp. 470–5, 480). Although cooperation can involve coincidence, coercion and/or coadjustment (Martin, 1992, pp. 25–7), cooperation in modern trade relations often revolves around the concept of reciprocity. Reciprocity, according to Carolyn Rhodes, 'refers to the maintenance of balance in trading relationships, where access to the domestic market is exchanged for access abroad and mutually agreeable rules of fair trade are established'. Part of the reciprocal relationship established is 'the expectation that when participants perceive imbalance or violate rules, then retaliation may be warranted to maintain balance' (Rhodes, 1993, p. 8). Prospects for cooperation in trade policymaking can decrease as the number of players involved in a trade negotiation increases.[11] Achieving cooperation in bilateral trade relations is an inherently different process from achieving cooperation in multilateral trade relations.[12]

The interaction of various negotiation strategies[13] can also affect a country's trade policy. Kishore Gawande and Wendy Hansen (1999) and others have studied the effectiveness of retaliation in lowering other countries' trade barriers.

Comparing countries' trade policymaking processes and possibilities for creating typologies of different countries' trade policymaking processes

The domestic and international factors discussed above are all interrelated and can be used in the analysis and comparison of the role of domestic politics and international relations in the trade policymaking of various countries, and in trade negotiations. By studying the relative importance in different countries of the domestic and international variables just discussed and comparing correlation across countries, empirical generalizations can be made about trade policymaking in different countries, and typologies that make distinctions between various forms of trade policymaking can be developed.[14] Countries can be grouped by the extent to which the administration of imports, the administration of exports and the conduct of trade negotiations are influenced by particular relationships between the state (bureaucracy), legislature and other actors in the trade policymaking process. Countries can also be grouped by the extent to which trade policy is affected by particular international factors such as the international trade regime, the position of the country in the international political economy, interdependence, the evolution of international capitalism, cooperation and negotiating strategies. Certain patterns in the relationship between domestic politics and international relations in trade policymaking might be found in countries with large bureaucracies devoted to the administration of either imports or exports. Certain patterns in the relationship between domestic politics and international relations in trade policymaking might also be found in countries where the legislature plays a more significant role in trade policymaking. Also, certain patterns in the relationship between domestic politics and international relations in trade policymaking might be found to be characteristic of certain time periods.

While the full elaboration of such typologies is beyond the scope of this book, this book does highlight some of the differences between trade policymaking in the world's two largest single-state economies – the United States and Japan – during a certain time period, 1986–1994, when a particular multilateral trade negotiation – the GATT Uruguay Round negotiations – occurred.

Developing and testing hypotheses that explore the relationship between domestic politics and international relations in a country during a particular trade negotiation

Given the complexity of trade relations between countries, there are a large number of hypotheses that could be designed and tested in order to explore the relationship between domestic politics and international relations in the trade policymaking of a particular country.

To limit the scope of inquiry, an arbitrary decision has been made in this book to examine the relationship between domestic politics and international relations in trade policymaking, by developing hypotheses that involve cooperation and the interaction of negotiating strategies at the international level in order to address the question: To what extent does domestic politics affect the agreement reached in an international trade negotiation? Such an inquiry follows along the lines of Martin's research concerning 'whether domestic institutions have any regular impact on patterns of international cooperation' (Martin, 2000, p. 192). However, the hypotheses developed in this book focus on more than just legislative-executive interaction at the domestic level, and emphasize cooperation as well as the interaction of negotiating strategies during trade negotiations at the international level, following Odell's suggestion that, 'We cannot understand international economic conflict and cooperation without a better grasp of the process of economic negotiation' (Odell, 2000, p. 2).

As Robert Mnookin *et al.* (2000, pp. 5–6) suggest, 'every legal negotiation involves a system of relationships', and in analyzing such negotiations, it is useful to conceptualize a negotiation as a 'system of relationships'. In a trade negotiation, the governments of the countries involved are negotiating between themselves concerning some trade matter. The legislature and other actors in either country may participate in the negotiations, either by exerting pressure on their own countries' negotiators, or by exerting pressure on actors in the other country, sometimes by forming transnational coalitions. To understand such a trade negotiation fully it is necessary to understand not only the domestic politics concerning the negotiation within each country, but also the relationship between the negotiators of the countries involved. Cultural factors, the history of past relations, the possibility of future relations, and conflict tendencies in the relations between two or more countries involved can affect the negotiation. Communication and information exchange problems, information asymmetries, role allocation and incentives can also affect the negotiation.

As Putnam suggests, the individual or individuals who negotiate for a particular country occupy a role between the domestic and international levels. A negotiator must often negotiate in two directions at the same time – with certain actors in the domestic politics of his or her own country, and with the negotiator, and sometimes other actors, from the other country (Putnam, 1988, p. 434).

The relationship between domestic politics and international relations in a trade negotiation changes during the various stages of a negotiation – some of which Odell (2000, pp. 26–8) outlines – the development of initial negotiating stances, the emergence of resistance points between the actors involved, the emergence of a zone of agreement between the actors involved, the reaching of an agreement, and finally discussions of implementing the agreement reached.

The following questions arise when one considers the role of a negotiator and the relationship between domestic politics and international relations in a particular trade negotiation:

- First, what roles are assigned to the negotiator, both in relation to the domestic politics of his or her country, and in relation to the other countries involved?
- Second, to what extent does domestic politics affect the negotiator's choice of tactics and strategies?
- Third, to what extent does domestic politics affect the agreement reached in the negotiation?

Using the conceptualization of the domestic level of trade policymaking discussed above and focusing on the variables of cooperation as well as the interaction of negotiating strategies at the international level in modern trade negotiations, this book addresses these questions by assessing the validity of the following three 'if, then' testable, falsifiable hypotheses, which are parallel to some of the hypotheses set out in Milner (1997), but do not assume, as does Milner (1997, pp. 33–4) that at the domestic level the political actors (executive and legislature) and societal actors (interest groups) are 'unitary and rational':

- First, if key elements within the state (bureaucracy) and the legislature of a country favor adopting a protectionist stance in a negotiation, then a government is more likely to adopt a protectionist negotiating stance, and is less likely to seek agreement concerning trade liberalization in an international trade negotiation.
- Second, the more division within a government concerning the negotiating stance that should be taken, the more influence the legislature, to the extent it has any influence, is likely to have over the terms of a negotiated agreement, and the more likely a government will be able to alter its negotiating stance if it chooses to cooperate in reaching an agreement.
- Third, if a substantial number of informed domestic groups endorse, or a key informed domestic group endorses, the negotiating stance of a government, then the legislature is more likely, if required, to endorse the results of the negotiation, and the government's chances of convincing other countries to cooperate in reaching an agreement increase.[15]

The structure of the rest of this book

Chapter 2 of this book demonstrates how the approach to analyzing trade policymaking and the hypotheses just described can be used to explore the relationship between domestic politics and international relations in the trade policymaking of the United States and Japan, and US–Japan trade negotiations. Chapter 2 concludes by discussing how the contextual two-level game approach to analyzing trade policymaking can be further developed by examining a particular case study – American trade policymaking, Japanese trade policymaking, and US–Japan trade negotiations during the GATT Uruguay Round agriculture negotiations.

Chapters 3, 4, and 5 test variations of the hypotheses described in Chapter 1 in order to analyze the relationship between domestic politics and international relations in the trade policymaking of the United States and Japan, and US–Japan trade negotiations during the GATT Uruguay Round agriculture negotiations.

Chapter 6 concludes by assessing the effectiveness of the contextual two-level game approach, and by suggesting areas for further research.

2
Domestic Politics and International Relations in US–Japan Trade Policymaking

Having discussed in Chapter 1 various domestic and international factors in trade policymaking, and having developed a series of hypotheses to examine the relationship between domestic politics and international relations during a particular trade negotiation, it is time to put such analytical tools to use, and to begin to explore the relationship between domestic politics and international relations in the trade policymaking of certain countries, during particular trade negotiations. This chapter demonstrates how the analytical tools described in Chapter 1 can be used to explore the relationship between domestic politics and international relations in the trade policymaking of the United States and Japan and in US–Japan trade negotiations.

Analyzing American trade policymaking

Diminishing domestic support for the United States government's efforts to promote international trade liberalization, and pressures for protectionism brought about by the large American trade deficit, have increased the need to understand the relationship between domestic politics and international relations in American trade policymaking.

Using the approach to analyzing trade policymaking discussed in Chapter 1, key domestic and international factors in the American trade policymaking process can be discerned, and specific hypotheses can be developed which can be used to analyze the relationship between domestic politics and international relations in American trade policymaking during a particular trade negotiation.

Existing domestic-oriented approaches to explaining American trade policymaking include state-centered, society-centered, ideas-oriented and rational choice approaches. State-centered approaches[1] to American trade policymaking generally argue that the United States is characterized by a weak state, which results from Congress having different political needs and constituencies from the Executive Branch, and results from private actors' especially powerful role in the American trade policymaking process. State-centered approaches to American trade policymaking, as described in Robert Pastor (1980, pp. 30–42, 49–60), generally emphasize the importance of either the Executive Branch or Congress, or the relationship between the two. More society-centered approaches to American trade policymaking emphasize the extent to which American trade policy is influenced by interest groups.[2] Some authors, such as Joanne Gowa (1988) and Michael Gilligan (1997), focus less directly on the state or society, and more on the relationship between state and society, emphasizing, for example, collective action. Ideas-oriented approaches examine the way in which ideas, such as liberalism or protectionism, have influenced American trade policymaking (Goldstein, 1993). Rational choice approaches analyze the extent to which American trade policymaking is a reflection of voters' preferences (Robert Baldwin, 1976, pp. 7–15; Charles Rowley and Willem Thorbecke, 1993, p. 349).

Using the conceptualization of the domestic level of trade policymaking described in Chapter 1, the domestic level of American trade policymaking can be conceived of as consisting of three key groups of actors – the Executive Branch (particularly the United States Trade Representative (USTR), the Departments of Commerce, Treasury, and State); Congress (especially the relevant Congressional committees); and other actors in the American trade policymaking process (including affected exporting and importing companies and concerned interest groups) – and the relationships between them.

Many diverse elements of the Executive Branch, Congress and other actors in the trade policymaking process, whose alliances may shift, participate in American trade policymaking. The trade policy that results, emerges in the context of shifting coalitions that form amid accommodation and conflict between the Executive Branch and Congress, which is at times heavily influenced by other actors in the trade policymaking process. The coalitions that form are affected by the policy preferences of the various actors involved, the nature of American political institutions, and the distribution of information among the various actors. These factors help to determine which elements of the Executive Branch and Congress, and which interest groups, are involved, what powers each has, and how their preferences evolve. The coalition building that occurs is characterized by America's adversarial legalism that favors the rights of the individual, more formal procedures (such as written agreements) and mutual suspicion (Thomas McCraw, 1986b, pp. 26–7). By analyzing these factors in a variety of trade disputes, patterns in the formation of coalitions between elements of the Executive Branch, Congress, and other actors in the trade policymaking process can be discerned.

The Executive Branch's trade policymaking authority derives from legislation passed by Congress that grants the Executive Branch the power to administer trade relations and to negotiate trade agreements. Within the Executive Office of the President, the USTR, the Cabinet-level coordinating group for economic policy, the National Security Council, the Council of Economic Advisers, and the Office of Management and Budget play important roles (Stephen Cohen *et al.*, 1996, pp. 109–10; US Congress, House Committee on Ways and Means, 1997, pp. 222–5). The USTR currently is the main trade negotiating authority, a role that was taken away from the State Department when the USTR's predecessor, the Special Trade Representative, was created in the Trade Expansion Act of 1962 (Cohen *et al.*, 1996, p. 109; Steve Dryden, 1995). During America's history, various interagency groups have been established in order to coordinate American trade policy (Cohen, 2000, pp. 71–96; US Congress, House Committee on Ways and Means, 1997, pp. 222–3).

The departments and agencies most involved in American trade policymaking include the Commerce Department, the Customs Service and the International Trade Commission (US Congress, House Committee on Ways and Means, 1997, pp. 225–8). The Departments of State, Treasury, Agriculture, Defense, and other departments and agencies become involved when certain issues are being addressed that concern them (Cohen *et al.*, 1996, pp. 110–14).

Divisions between the Executive Office of the President, and between and among departments and agencies, and within interagency coordinating groups, affect American trade policy. While many of the divisions within the Executive Branch are not visible to the public, differences between and within departments concerning trade policymaking sometimes become public during Congressional hearings. The Executive Branch's role in American trade policymaking has recently increased (Cohen, 2000, pp. 45–70; I. M. Destler, 1995, pp. 105–37) partly as a result of the growth of the administrative trade remedy system, which is designed to protect American industries against unfair trade practices. American trade policymaking is also affected by relations between the federal government and the states (Beaumont, 1996).

The laws and institutional arrangements that govern American trade policymaking grant Congress a particularly active role in the creation, implementation and oversight of American trade policymaking (US Congress, House Committee on Ways and Means, 1997, pp. 221–2; Cohen *et al.*, 1996, pp. 114–16; Destler, 1995, pp. 65–104). Congress has the constitutional authority to 'regulate commerce with foreign nations' and to 'lay and collect taxes, duties . . . and excises' (US Constitution, Art. I § 8; see Jackson *et al.*, 1995, pp. 112–16). Congress has the power to specify where and how appropriations are to be spent, to use oversight hearings, and to produce detailed committee reports concerning new trade legislation in order to clarify congressional intent, as well as to veto some actions by executive agencies that concern trade. Trade legislation, such as the Reciprocal Trade Agreements Act of 1934 (see Goldstein, 1993, pp. 152–4), the Trade Expansion Act of 1962, the Trade Act of 1974 (see Pastor, 1980, pp. 136–85), the Trade Agreements Act of 1979 (see Jackson, 1984), the Omnibus Trade and Competitiveness Act of 1988, and the Uruguay Round Agreements Act of 1994 (see David Leebron, 1997), can have far-reaching effects on American trade policymaking (Destler, 1994).

Congress has at times granted the authority to the President to negotiate reciprocal trade agreements that affect American laws, but has also required the transmission of such trade agreements to Congress. Congress has done this more recently by granting the President fast-track negotiating authority or trade promotion authority, which makes approval of a trade agreement negotiated by the President subject to a procedure in which Congress cannot amend the agreement reached.

Congressional power concerning trade policymaking is often centered in particular congressional committees and their chairs, most notably the Senate Committee on Finance and the Committee on House Ways and Means, as well as in committees concerned with particular trade issues.

Congress's role in trade policymaking is often affected by interest groups that have varying degrees of access to Congress. Because trade policy can become highly politicized and be portrayed as having a direct impact on employment and profits, trade policymaking can be marked by active interest group lobbying, which some argue can lead to more distributive, narrower and short-term policies.

The Republican and Democratic parties, the two major postwar parties in America's single-member plurality district electoral system, have shared a postwar bipartisan consensus concerning trade policy. This consensus has recently been weakened as a result of record trade deficits, and as a result of growing concern over the impact of international trade liberalization on labor and the environment (Raj Bhala, 1996, pp. 1183–360).

The pattern of accommodation and conflict between the Executive Branch and Congress is affected by the policy priorities of the Executive Branch and the Congress and by the political parties (Pastor, 1980, pp. 61, 186–99; Pastor, 1983, p. 188). Divided party control in which one party controls the Presidency and the other party controls the Congress, or in which different parties control the House of Representatives and the Senate, can affect trade policymaking.

Various other actors play a significant role in the American trade policymaking process. Interest groups play a key role in pressuring to have certain issues placed on the government's trade policymaking agenda, and in shaping trade policy concerning these issues (Pastor, 1980, pp. 43–9; Cohen, 2000, pp. 113–36). Interest groups can provide government agencies and the Congress with detailed information concerning business conditions in certain sectors and the behavior of certain firms. Private-sector advisory committees that were established in the Trade Act of 1974 provide another avenue of influence for interest groups (US Congress, House Committee on Ways and Means, 1997, pp. 228–9). The close working relationship that has evolved between the Executive Branch and interest groups is particularly evident in the trade remedy system, in which an American producer or association representing an industry or group of workers may request the United States government to initiate an antidumping or countervailing duty investigation in order to determine whether an imported product is being sold in the United States at less then fair value, or is being improperly subsidized and is causing material injury to an American industry (Bhala, 1996, pp. 601–865; Cohen *et al.*, 1996, pp. 141–61; Destler, 1995, pp. 139–73; Jackson *et al.*, 1995, pp. 666–814).

Many international factors also influence American trade policy, as is outlined in the next few paragraphs.

The United States has agreed to have certain constraints placed on its trade policy as part of the mutual trade agreements that the country has entered into with the other member countries of the GATT and the WTO. The GATT and the WTO have served the United States as instruments for regime creation and maintenance, for institutionalizing dispute settlement procedures, for managing protectionist pressures, and for legitimizing foreign policy objectives (Margaret Karns, 1990, pp. 142–61). At the regional level, the United States has entered into similar mutual obligations with Canada and Mexico under the North American Free Trade Agreement (NAFTA).

During the postwar period, the United States has often acted as a dominant or hegemonic state in the postwar international political economy, establishing and maintaining the international trade regime institutionalized in the GATT and the WTO through providing collect-ive goods as well as by coercing reluctant states (Robert Gilpin, 1987, pp. 85–92, 343–60; Ikenberry *et al.*, 1988b, pp. 4–5; Keohane, 1980, 1984; Duncan Snidal, 1985; Arthur Stein, 1984). Recently, the United States has adopted a more aggressive trade policy in economic sectors where it is losing market share.

Growing interdependence, which is partially the result of the United States' promotion of multilateral international trade liberalization during the postwar period, increasingly has constrained the American government's freedom of action in designing its trade policy.

The evolution of capitalism (Immanuel Wallerstein, 1989, pp. 193–256), the period during which the United States industrialized (Barring-ton Moore, 1966, pp. 111–55), and changes in the trade policymaking of other countries and within the GATT or the WTO affect American trade policy.

The United States has sought to design and implement a trade policy in order to realize absolute or relative gains, maintain friendly rela-tions, and in recognition of power asymmetries. Prospects for United States' cooperation are affected by the number of other countries in-volved in a particular trade negotiation (Jackson, 1987, p. 395).

The interaction of particular negotiating strategies also affects American trade policy (Schoppa, 1997, pp. 18–48).

Various phases in the relationship between domestic politics and international relations in American trade policymaking

By analyzing the evolving relationship between the domestic and international factors just outlined, various phases in the relationship between domestic politics and international relations in American trade policymaking can be discerned, although no more than a very brief outline of these phases is possible in this book.

Between 1789 and the Civil War (1861–5), Congress played a domin-ant role in American trade policymaking. The first Act of Congress in 1789 was to establish a revenue tariff of about 8.5 per cent *ad valorem* (Alfred Eckes, 1999, p. 58). During most of the nineteenth century, northern manufacturers exerted pressure for a higher tariff, while Southern agricultural producers argued for a lower tariff (Eckes, 1995, pp. 1–27; Goldstein, 1993, pp. 23–80). While a protective tariff was designed to promote national economic development, it was not popular with exporters. The United States was mostly concerned with expanding exports to Europe and the Caribbean, and with the frontier (Wallerstein, 1989, p. 227). Until the 1890s, the United States' major trading partner was Britain (Eckes, 1999, p. 52).

During the late nineteenth and early twentieth centuries, as David Lake argues, domestic political pressures provided 'the best explanation of the pattern of protection across industries and the specific rate of duty established for each industry'. Despite the United States' 'low level of dependence on the international economy, large domestic market, isolationist ideology, and permeable political process dominated by domestic pressures, the trade strategy of the United States was influenced in important ways by the structure of the international economy' (Lake, 1988, pp. 216, 227). Congress, which was mostly under the control of the Republican Party, continually raised levels of tariff protection, responding to producer pressures, especially from western farmers, eastern manufacturers, and big business. A more complex administrative structure to handle trade policymaking emerged, as the Office of the Commissioner of Revenue was created, followed by the Bureau of Foreign and Domestic Commerce. Increasing pressure was placed on the government to expand access to foreign markets, and to protect domestic industry from foreign competitors in the United States who were selling goods at lower prices than in their own countries. The President was given the right to negotiate reciprocal trade agreements in the Tariff Act of 1913, subject to Congressional ratification. Presidential control over tariffs increased during the 1920s with the Fordney–McCumber Act. Antidumping legislation was passed in 1916 and 1921. The House Committee on Ways and Means became increasingly important in Congressional trade policy, as a more complicated tariff structure emerged. The Republican Party, backed by agricultural interests, favored a higher tariff, while Democrats favored a lower one. As agricultural prices fell dramatically in the 1920s, pressures for a higher tariff increased, resulting in the Smoot-Hawley Tariff Act of 1930, which raised the average rate on dutiable imports to near 50 per cent (Goldstein, 1993, pp. 81–136; Gourevitch, 1986, pp. 105–11, 147–53). The United States became increasingly important in the international political economy as British hegemony declined.

The extreme protectionist tariffs of the Smoot-Hawley Act were one of the main reasons for the Great Depression. Congress, beginning with the Reciprocal Trade Agreements Act of 1934, sought to reduce the direct influence of producers on Congress, and transferred some of Congress's trade policymaking responsibilities to the Executive Branch. The President was permitted to lower tariffs during a specified time period, and to set individual tariffs, as well as to implement, through executive order, treaties that the Executive Branch had negotiated involving tariff reductions. The Reciprocal Trade Agreements Act of 1934 centered control over trade policymaking in the State Department, and encouraged the Executive Branch to sign trade agreements concerning manufactured products that were based on the concepts of reciprocity and most-favored nation status. But signing such trade agreements concerning agricultural products was not favored by American agricultural interest groups, and the Agricultural Adjustment Act of 1933, as amended in 1935, permitted the President to impose import quotas and use export subsidies in order to protect American agriculture (Goldstein, 1993, pp. 146–57).

A fragmented Executive Branch, a decentralized Congress, and societal actors who have easily been able to influence trade policymaking have marked postwar American trade policymaking (Krasner, 1978a, pp. 57–71). The State Department pursued a liberal strategy towards manufacturing that emphasized market forces and the reduction of trade barriers. As a result of this strategy, the United States government developed relatively few policy instruments that could be applied directly to particular sectors or firms, and provided less support than many countries for selected companies engaged in international trade (David Yoffie, 1986, p. 38). Congress generally supported the Executive Branch's trade policy while acknowledging calls for protectionism from affected industries (Robert Baldwin, 1976, pp. 15–37). The United States sought to strengthen the international trade regime through the GATT. Although the original members of the GATT were interested in eliminating quantitative restrictions when they first met to draft the General Agreement on Tariffs and Trade in the late 1940s, by 1955, the United States, under pressure from agricultural interest groups and Congress, requested and received a waiver for the agricultural import quotas that the American government had established pursuant to the Agricultural Adjustment Act of 1933. Partly because the United States abandoned support for the International Trade Organization that was designed to accompany the GATT treaty (Goldstein, 1993, pp. 157–63), the International Trade Organization did not come into being (Susan Aaronson, 1996, pp. 61–132). Nevertheless, the GATT evolved into a form of international trade organization that came to codify trading relations among its member nations.

During the 1960s and 1970s, the growth of the American administrative state led to the creation of new offices with international trade responsibilities, such as the Special Trade Representative. Smaller offices in various departments that were concerned with foreign economic policy expanded into full-scale bureaus, and a more complex trade remedy system emerged. Strong Congressional committees worked with the Executive Branch in order to support the creation and maintenance of a liberal international order (Destler, 1995, pp. 27–30). This policy, Katzenstein notes, 'ran parallel to and was reinforced by the economic interests of the business community' (Katzenstein, 1978b, p. 308). Congress became increasingly assertive, but more dispersed, as public access to the legislative process increased. Several firms in industries that were affected adversely by imports sought protection (Milner, 1988, pp. 103–58; 222–63). There was an explosion in international trade, a rise in new competitors, and increasing trade with Asia (Destler, 1995, pp. 41–63). The international trade regime became increasingly institutionalized as a result of the Dillon, Kennedy and Tokyo Rounds of GATT negotiations, in which the United States played a major role.

During the early 1980s, the Executive Branch continued to pressure for international trade liberalization, but record unemployment, an overvalued dollar and an increasingly large trade deficit fueled protectionist pressures on Congress. Exporters called on the American government to assist them in gaining access to foreign markets. Heightened Congressional concern over the domestic effects of international trade liberalization made it increasingly difficult for the Executive Branch to secure the extension of trade negotiating authority. The United States played a major role in the Uruguay Round negotiations, which lasted from 1986 to 1994, and led to the creation of the WTO.

Although the economic boom of the mid- to late 1990s resulted in less concerns being voiced about the size of the American trade deficit, growing complaints about the effects of international trade liberalization on labor and the environment led Congress to deny trade negotiating authority to the Clinton Administration after the Uruguay Round negotiations, and led to widespread demonstrations when efforts were made to launch a new negotiating round at the December 1999 WTO Ministerial Meeting in Seattle.

Since then, the United States has been actively involved in the new WTO Doha Agenda negotiations launched at the November 2001 meeting in Doha, Qatar. In August 2002, the US Congress granted the President trade promotion authority to negotiate trade agreements.

Designing and testing hypotheses that explore the relationship between domestic politics and international relations in American trade policymaking during a particular trade negotiation

Given the complexity of American trade relations, there are a large number of hypotheses that could be designed and tested in order to explore the relationship between domestic politics and international relations in American trade policymaking. In order to limit the scope of the inquiry, this book explores the relationship between domestic politics and international relations by developing hypotheses that involve cooperation and the interaction of negotiating strategies at the international level.

The following questions arise when one considers the role of an American negotiator and the relationship between domestic politics and the agreement reached in a particular trade negotiation:

- First, what roles are assigned to the American negotiator, in relation to American domestic politics and to the other countries involved?
- Second, to what extent does American domestic politics affect the American negotiator's choice of tactics and strategies?
- Third, to what extent does American domestic politics affect the agreement the United States reaches in a trade negotiation?

Using the conceptualization of the domestic level of American trade policymaking described above, and focusing on cooperation and the interaction of negotiating strategies at the international level in modern American trade negotiations, this book addresses the above questions by assessing the validity of the following hypotheses, which are derived from the hypotheses listed in Chapter 1:

- First, if key actors within the Executive Branch and the Congress favor a protectionist stance in a negotiation, then the American government is more likely to adopt a protectionist negotiating stance, and is less likely to seek trade liberalization in an international trade negotiation.
- Second, the more division there is within the American government concerning the negotiating stance that should be taken, the more influence Congress, if it has any influence, is likely to have over the terms of the negotiated agreement, and the more likely the American government will be to alter its negotiating stance if it chooses to cooperate in reaching an agreement.

- Third, if a substantial number of informed domestic groups endorse, or a key informed domestic group endorses, the American government's negotiating stance, then Congressional endorsement of the results of the negotiation (if required) is more likely, and the American government's chances of convincing other countries to cooperate in reaching an agreement increase.

Analyzing Japanese trade policymaking

The recent emergence of Japan as the second-largest economy in the world, and the growing number of trade disputes involving Japan have placed increasing pressure on actors in the Japanese trade policymaking process to conform with GATT and WTO rules and increased the need to understand the relationship between domestic politics and international relations in Japanese trade policymaking.

Using the approach to analyzing trade policymaking outlined in Chapter 1, key domestic and international factors in the Japanese trade policymaking process can be discerned, and specific hypotheses developed which can be used to analyze the relationship between domestic politics and international relations in Japanese trade policymaking during a particular trade negotiation.

The domestic level of Japanese trade policymaking can be analyzed using state-centered, society-centered, ideas-oriented, and rational choice approaches. State-centered approaches generally emphasize the important role that the Japanese state plays in Japanese trade policymaking, and are linked to theories of bureaucratic dominance in Japanese policymaking (Chalmers Johnson, 1982, pp. 17–34, 305–24). Society-centered approaches to Japanese trade policymaking focus on the role that various interest groups play in Japanese trade policymaking. Such theories often bear some similarities to the tripartite elite model of Japanese policymaking (Haruhiro Fukui, 1977, pp. 22–35). One well-known conceptualization of Japanese foreign economic policy from the 1970s characterizes Japanese foreign economic policy as involving a 'corporatist coalition of finance, major industry, trading companies, and the upper levels of the national bureaucracy, coupled with the consistent rule of the conservative Liberal Democratic Party, [and] the systematic exclusion of organized labor from formal policy-making channels' (T. J. Pempel, 1978, p. 139). Daniel Okimoto similarly emphasizes the major role played in the 1970s and 1980s by the dominant Liberal Democratic Party's 'grand coalition of interest support, with its bureaucratic division of labor and issue-specific power configurations that form[ed] dynamically into separate, semi-self-contained policy domains' (Okimoto, 1988, p. 340). Some advocates of society-centered approaches to Japanese trade policymaking emphasize the important role of industry actors in compromising or permeating the Japanese bureaucracy (Uriu, 1996, p. 8). Ideas-oriented approaches often emphasize the role that neo-mercantilist ideas play in Japanese trade policy (Kenneth Pyle, 1996, pp. 36–9), as reflected in the government's active intervention in the market, and in the large number of instruments Japanese policymakers have been able to use to affect specific sectors and firms directly (Okimoto, 1989, pp. 24–7). However, some argue that Japan has recently adopted more of a free-trade philosophy (Ryutaro Komiya and Motoshige Itoh, 1988, pp. 205–6). A rational choice approach to analyzing Japanese trade policymaking would argue that Japanese trade policymaking is a reflection of firms seeking to maximize profits, and government officials trying to stay in office. Such arguments might be based on J. Mark Ramseyer and Frances Rosenbluth, 1995 (see also Ramseyer and Rosenbluth, 1993).

Following the conceptualization of the domestic level of trade policymaking described in Chapter 1, the domestic level of Japanese trade policymaking can be conceived of consisting of three key groups of actors – the bureaucracy (particularly the Ministry of Economy, Trade and Industry (formerly the Ministry of International Trade and Industry) and other ministries concerned with a particular trade issue); the Diet (including the major parties such as the Liberal Democratic Party (LDP)); and other actors in the Japanese trade policymaking process (including importing and exporting companies, and concerned interest groups) – and the relationships between them.

Many diverse elements of the bureaucracy, the Diet and other actors in the trade policymaking process, whose alliances may shift, participate in Japanese trade policymaking. The trade policy that results, emerges in the context of shifting coalitions that form as the bureaucracy responds to pressures exerted on it by elements within the bureaucracy, the Prime Minister's Office, the Diet, and interest groups. The shifting coalitions that occur form within the framework of the bureaucracy, which structures procedurally the types of possible alliances and policymaking patterns of interest groups, and the party system in the Japanese Diet.[3] The coalitions that form are affected by the policy preferences of the various actors involved, the nature of Japanese political institutions, and the distribution of information among the various actors. These factors help to determine which elements of the bureaucracy, which Diet members, and which interest groups are involved, what powers each has, and how their preferences will evolve. The coalition building that occurs is characterized by more concern for group welfare, informal procedures, oral agreements and mutual trust, than is the coalition building that occurs in American trade policymaking (McCraw, 1986b, pp. 26–8). By analyzing these factors in a variety of trade disputes, patterns in the formation of coalitions between elements of the bureaucracy, the Diet and other actors in the trade policymaking process can be discerned.[4]

The bureaucracy is often the most important actor in Japanese trade policymaking, mediating between domestic political forces, and coordinating the overall trade policymaking process (Okimoto, 1988, p. 318). The rigid separation of powers between the Executive Branch and the Congress that exists in the United States does not exist in Japan's parliamentary system. Since the Prime Minister can appoint only the Ministers and Vice Ministers of the various Ministries, and because Japan's bureaucracy is more merit-based and smaller than the American bureaucracy, Japanese trade policymaking is marked by fewer high-level political appointees than in America (Destler *et al.*, 1976, p. 86).

During the postwar period, the key players in Japanese trade policy-making within the bureaucracy have generally been the Ministry of International Trade and Industry (Chalmers Johnson, 1977) (now the Ministry of Economy, Trade and Industry) (which has coordinated trade policy) and the Ministry of Finance (which has overseen policies concerning government finance and expenditures), although the Ministry of Finance's power has been reduced in recent years, in an effort to restructure the Japanese bureaucracy. Many argue that the Japanese government has made greater efforts than its American counterpart to coordinate economic units within sectors, and to plan overall sectoral economic strategies. The Ministry of Foreign Affairs has been described as being '[f]airly influential (in terms of broad involvement)' in trade policy, while the Ministry of Agriculture, Forestry and Fisheries, the Ministry of Posts and Telecommunications (now the Ministry of Public Management, Home Affairs, Post and Telecommunications), and the Ministry of Health and Welfare (now the Ministry of Health, Labour and Welfare) have exerted a lot of influence when trade issues have arisen that concern them. The Ministry of Local Autonomy (recently merged into the Ministry of Public Management, Home Affairs, Post and Telecommunications) has been 'very powerful' concerning local budget issues. In contrast, the Ministry of Labour (recently merged into the Ministry of Health, Labour and Welfare), the Defense Agency, the Science and Technology Agency (recently merged with the Ministry of Education), and the Economic Planning Agency (recently merged into the Cabinet Office) have been not as influential in Japanese trade policymaking.[5]

Tension and conflict can emerge among the various agencies and ministries involved in formulating and executing Japanese trade policy. Mitsuo Matsushita notes that this tension is 'something comparable to the conflict and tension between the Congress and the Executive Branch in the United States'. However, ministerial conflicts in Japan are much less public, and 'are not openly reported . . . even though they may be common knowledge among those who know something about the Japanese government'. Matsushita states that 'No official report or comments are made. Therefore, we must be satisfied with some reports that have trickled out through newspaper accounts' (Matsushita, 1984, p. 82).

The governing coalition within the Diet, which at times during the postwar period has included only the LDP, chooses the Prime Minister. Trade policymaking sometimes involves tensions that arise within the bureaucracy, particularly between the Ministry of Economy, Trade and Industry and other ministries, which can require a delicate balancing act by the Prime Minister (Kenji Hayao, 1993) and key parliamentary leaders. Cabinet officials are generally chosen from the various political parties or factions within the governing coalition based on the balance of parties or factions within the Diet, rather than being selected personally by the chief of government, as in the United States. Cabinet members involved in trade policymaking are often aligned more closely to certain Diet members or factions than American Cabinet members are to particular groups within the US Congress.

The Diet's role in Japanese trade policymaking is influenced by the Japanese Constitution and the configuration of political parties within the Diet. Article 73(iii) of the 1947 Constitution states that the Cabinet shall conclude treaties, however, Article 73(iii) notes that the Cabinet 'shall obtain prior or, depending on circumstances, subsequent approval of the Diet'.[6] L. Jerold Adams states that not only 'constitutional provisions state that Diet approval is requisite for the validity of ratification', but 'the factional nature of the political order also constitutes a check on the treaty-making powers of the Cabinet' (Adams, 1974, pp. 33–4). But there are a number of forms of international trade agreements that are not considered as needing the approval of the Diet according to Matsushita (1993a, pp. 27–9). Since the mid-1950s, while five or six parties have routinely won seats in the Diet, the LDP has consistently held a large number of them (Curtis, 1988, pp. 1–44). When the LDP has been dominant in the Diet, it has exerted substantial influence over trade policymaking, particularly through groups of Diet members, sometimes called policy tribes of Diet members (*zoku*), who are knowledgeable and adapt at pressuring certain ministries about particular issues (Curtis, 1999, pp. 53–5). Such groups have often included former bureaucrats who became LDP Diet members. Okimoto argues that, during the 1970s and 1980s, as a result of the length of time that the LDP was dominant, its 'constitutional authority and electoral mandate' were blended with 'the bureaucracy's administrative skills and technical know-how to such an extent that the relative power of the ministries in trade policymaking waned (Okimoto, 1988, pp. 319, 326). During this period, 'a legislative proposal made by the Cabinet after extraparliamentary consent by the LDP [was] likely to pass'. The opposition parties generally did not object to proposed bills in the area of foreign and international economic policy, 'possibly because they were too complex and technical and not very interesting' to them (Matsushita, 1984, p. 80). Also, as minority parties, they were not easily able to initiate or block important legislation. Furthermore, most of the population was profiting from Japan's rapidly expanding economy.

As trade tensions rose and LDP dominance declined from the late 1980s to the mid-1990s, Diet members became much more vocal concerning trade policymaking, as Japan was forced to make concessions in a growing number of trade disputes, many of which involved agriculture. These were of direct relevance to many Diet members, whose constituents might be affected by the concessions under discussion. As will be discussed in Chapter 4, partly as a result of such trade disputes, fundamental divisions emerged within the LDP. The party was unable to maintain a majority in both Houses of the Diet, and thus came to play a diminished role in the trade policymaking process.

Interest groups often play a less visible role in the Japanese trade policymaking process than in the same process in America, partly because the frequent and highly influential Congressional hearings that occur in American trade policymaking do not exist in Japan. Large trading companies carry out much of Japanese trade, and interest groups are often more allied within a sector, because of the particular pattern of intermediate organizations and associations that has been established in Japan (Okimoto, 1988, pp. 313–16). Muramatsu and Krauss (1987) and Okimoto (1988, p. 308) argue that recently interest groups have become stronger and 'relatively autonomous from the state', although this may just reflect a breakdown in the longstanding consensus that existed from the 1950s to the 1980s concerning Japanese economic policy (Pempel, 1993, pp. 123–30).

Many of the international factors mentioned in Chapter 1 affect Japanese trade policymaking, as outlined in the next few paragraphs.

Japan has agreed to have certain constraints placed on its trade policy as part of the mutual trade agreements that Japan has entered into with other countries. Japan became a member of the GATT during the 1950s (Tatsuo Akaneya, 1992). By the 1970s, it was one of the most active participants in the GATT Tokyo Round negotiations (Komiya and Motoshige Itoh, 1988, pp. 203–5). As Japan emerged as the second largest economy in the world during the 1970s and 1980s, and as various countries sought to increase their imports to Japan as their trade deficits with the country escalated, pressure on Japan to open its markets increased (Gilpin, 1988).

While Japan's postwar success in international trade led to discussions of 'Japan's potential as an economic hegemon or at least co-equal partner with the United States for international economic leadership' (Philip Meeks, 1993, p. 58), a number of '[c]onfining international conditions' continue to affect Japan's trade policy. 'Japan's international geographic position is weak.' Japan's 'domestic natural resource capabilities' are 'virtually nil'; and during the postwar period, Japan has had 'an extremely high strategic dependence on a single country, the United States' (Pempel, 1978, p. 142). See also Nobutoshi Akao (1983).

Growing interdependence has resulted in a large number of trade disputes involving Japan which increasingly have forced actors in the Japanese trade policymaking process to conform to GATT and WTO rules. In response, Japan has used the WTO dispute settlement system more actively to protect its interests.

The evolution of capitalism, the period during which Japan industrialized, and changes in other countries' trade policymaking – as well as changes within the GATT and the WTO – have affected Japanese trade policy. Certain elements of Japanese trade policymaking have been adopted from other countries, or in response to changing practices in the international trade regime.

Japan has sought to design and implement a trade policy in order to realize absolute or relative gains, to maintain friendly relations, and in recognition of power asymmetries. Prospects for Japan being able to cooperate are affected by the number of other countries involved in a particular trade negotiation.

The interaction of particular negotiating strategies affects Japanese trade policy. Some authors have attempted to identify particular characteristics of the Japanese negotiating style (Michael Blaker, 1977a, 1977b; John Graham, 1993). Mayumi Itoh suggests that Japanese protectionism results from an exclusionist mind-set (*sakoku* mentality) which 'constitutes the core of Japanese foreign policy decision makers' attitudinal prism' (Mayumi Itoh, 1998, p. 14). Kent Calder, on the other hand, characterizes Japanese foreign economic policy as reactive or passive, responding to 'outside pressures for change albeit erratically, unsystematically, and often incompletely' (Calder, 1988a, p. 519). Others argue that since the GATT Tokyo Round negotiations of the 1970s, Japan has played a particularly active role in international trade relations (Komiya and Motoshige Itoh, 1988, pp. 203–5).

Various phases in the relationship between domestic politics and international relations in Japanese trade policymaking

By analyzing the evolving relationship between the domestic and international factors just outlined, various phases in the relationship between domestic politics and international relations in Japanese trade policymaking can be discerned, although no more than a very brief outline of these phases is possible in this book. Given the focus of this book on modern trade relations, only the last few hundred years of Japan's long history is considered in the paragraphs that follow.

From the mid-seventeenth to the mid-nineteenth centuries, the Tokugawa Shogunate controlled trade relations strictly with the outside world. Copper was exported in exchange for cotton and silk textiles, spices, tea, medicines and sugar (Christopher Howe, 1996, pp. 37–41). However, 'this trade had relatively little effect on the Japanese economy as a whole' (John Fairbank *et al.*, 1989, p. 409).

The opening of Japan to trade with other countries by the United States and a number of European countries in the mid-nineteenth century 'precipitated a crisis in relations between the bakufu [shogunal government] and the domains' (E. Sydney Crawcour, 1997, p. 36), and to the signing of what came to be known as the 'unequal treaties'. These treaties limited Japan's tariff rates until they were later revised. During each decade of the Meiji era (1858–1911), Japan's shipments overseas 'doubled in volume, more or less' (William Lockwood, 1968, pp. 539, 336). During the early Meiji period, efforts were made to reduce foreign merchants' domination of Japan's foreign trade (Arthur Tiedemann, 1974, pp. 129–37). An active debate occurred between those who emphasized the natural self-sufficiency of Japan and those

who took a more nationalistic approach to foreign trade. Japan's legislature, the Diet, was created in the 1889 Constitution. Control over
domestic and foreign commercial policy came to be centered in the
Ministry of Agriculture and Commerce, and the government inter-
vened directly in the economy in order to promote certain enterprises
(Howe, 1996, pp. 100–11).

After having gained control of some of its own tariffs, the Japanese
government adopted policies including 'subsidies to the Japanese ship-
building industry and shipping lines, tariff and tax measures to promote trade with Japan's colonies, tariffs and other policies to restrict
importation of many consumer and producer goods', as well as an
exchange rate policy designed 'to boost exports and minimize the
trade deficit' (Kozo Yamamura, 1995, p. 115 see also Howe, 1996, pp.
164, 292–315). A large and important business community evolved
which strove 'to reach a consensus on trade issues and to discuss them
regularly with officials who, in turn, became increasingly responsive in
playing a mediating role to help specific industries cope with problems
and in composing new laws' (William Fletcher, 1989, pp. 6–7). During
the 1920s, partially as a result of tensions between agriculturists and
industrialists, the Ministry of Agriculture and Commerce split into the
Ministry of Agriculture and Forestry, and the Ministry of Commerce
and Industry (Chalmers Johnson, 1982, pp. 83–95).

During Japan's Pacific War, which included the Second World War,
the government sought to secure access to key resources and to promote textile and manufacturing exports, particularly to its colonies, as
the government came to exert increasing control over the Japanese
economy, and as the world economy entered a depression and international trade slowed (Fairbank *et al.*, 1989, pp. 654–5; Takafusa
Nakamura, 1997, pp. 147–57). After the war, during the American
Occupation of Japan, Japan's trade relations with the world were
strictly curtailed.

Even though the United States tried to dismantle Japan's large
trading companies during the early years of the American Occupation,
after the Occupation ended, the Japanese bureaucracy, led by the
Ministry of International Trade and Industry, continued the highly
successful prewar policy of promoting exports in high-priority industries and restraining some imports, such as consumption goods, and
using selective trade measures (Komiya and Motoshige Itoh, 1988,
pp. 175–7; Matsushita, 1993a, pp. 170–295). A conservative coalition
emerged as a two-party system developed within Japan's reconstructed
democracy. Business groups supported a trade policy that would reduce

unemployment, promote Japan's economic development, and lead Japan to be economically independent.

The domestic consensus that emerged in favor of such policies helped to support the longstanding postwar rule of a conservative elite that crystallized in the LDP's dominance of the Diet. The success of Japan's economic development muted the influence of domestic political debate on Japan's trade policy (Donald Hellmann, 1988, pp. 346, 356–8). Japan reasserted itself into the world economy, becoming a member of the GATT in 1955, and replaced many import controls with tariffs during the early 1960s (Toru Nakakita, 1993, pp. 346–50). Reluctantly participating in the GATT Kennedy Round negotiations (1963–7), Japan agreed to a series of tariff reductions and the liberalization of imports, not so much in order to improve the allocation of resources and promote competition, but as 'a necessary sacrifice for Japan to become a member of the international economic community' (Komiya and Motoshige Itoh, 1988, pp. 178–86).

During the mid-1960s to the mid-1970s, as the Japanese economy expanded rapidly, and as the LDP's rule continued, the bureaucracy and the LDP came to work more closely together on trade policy, focusing on the promotion of exports. As Japan transformed itself from 'a relatively small country exporting primarily labor-intensive goods' to a large and important 'actor in the world economy exporting largely sophisticated machinery', increasing tension emerged between Japan's traditional neo-mercantilist trade policies and the requirements of Japan's membership in the GATT. Mostly in response to foreign pressure, Japan further reduced its formal import restrictions on manufactures and agricultural products (Komiya and Motoshige Itoh, 1988, pp. 173, 185–91).

During the 1970s, while the LDP transformed itself into a catch-all party in order to maintain its dominance (Takashi Inoguchi, 1990) and became 'far less corporatist and far more pluralistic in terms of the interest groups it [was] willing to work with' (Pempel, 1993, p. 127), key elements of the support base of the LDP continued to be located in relatively inefficient, domestic, protected economic sectors, such as agriculture, retail distribution and public works. A balance needed to be found between those who favored freer trade and those who sought to maintain the status quo (Komiya and Motoshige Itoh, 1988, p. 210). By the time of the GATT Tokyo Round negotiations (1973–9), Japan had assumed a much larger percentage of world industrial production and trade and took on a greater role in the negotiations (Mark Manyin, 1999, pp. 111–62). As a result of the negotiations, Japan agreed to a substantial number of tariff reductions, and to various agreements concerning the reduction of non-tariff barriers, which resulted in Japan's average rate of tariffs reaching a relatively low average level in comparison to other developed countries.

By the mid-1980s, efforts at domestic deregulation were often linked to international trade negotiations. From the mid-1980s to the late 1990s, as discussed in Chapter 4, LDP dominance declined (Ray Christensen, 2000; Curtis, 1999) as the coalition of business and agricultural groups that had long supported the LDP unraveled. Japan participated actively in the GATT Uruguay Round negotiations (1986–94), staunchly resisting the liberalization of its rice market while seeking to change elements of the international trade regime that Japan might be able to use to its advantage, such as the rules concerning antidumping (Manyin, 1999, pp. 163–312).

Under the WTO, and during the WTO Doha Agenda negotiations, as Japan's economy has faltered, the country has resorted increasingly to the new WTO dispute settlement system (Amy Searight, 2002, pp. 174–9).

Designing and testing hypotheses that explore the relationship between domestic politics and international relations in Japanese trade policymaking during a particular trade negotiation

Given the complexity of Japanese trade relations, there are a large number of hypotheses that could be designed and tested in order to explore the relationship between domestic politics and international relations in Japanese trade policymaking. In order to limit the scope of the inquiry, this book explores the relationship between domestic politics and international relations by developing and testing hypotheses that involve cooperation and the interaction of negotiating strategies at the international level.

The following questions arise when one considers the role of a Japanese negotiator and the relationship between domestic politics and the agreement reached in a particular trade negotiation:

- First, what roles are assigned the Japanese negotiator, both in relation to Japanese domestic politics and to the other countries involved?
- Second, to what extent does Japanese domestic politics affect the Japanese negotiator's choice of tactics and strategies?
- Third, to what extent does Japanese domestic politics affect the agreement Japan reaches in a trade negotiation?

Using the conceptualization of the domestic level of Japanese trade policymaking described above, and focusing on cooperation and the interaction of negotiating strategies at the international level in modern Japanese trade negotiations, the chapters that follow address the above questions by assessing the validity of the following hypotheses, derived from the hypotheses at the end of Chapter 1:

- First, if key actors within the bureaucracy and the Diet favor the Japanese government adopting a protectionist stance in a negotiation, then the Japanese government is more likely to adopt a protectionist negotiating stance, and is less likely to seek agreement concerning trade liberalization in an international trade negotiation.
- Second, the more divided the Japanese government is concerning the negotiating stance that should be taken, the more influence the Diet, to the extent that it has any influence, is likely to have over the terms of the negotiated agreement, and the more likely the Japanese government will be able to alter its negotiating stance if the Japanese government chooses to cooperate in reaching an agreement.
- Third, if a substantial number of informed domestic groups endorse, or a key informed domestic group endorses, the Japanese government's negotiating stance, then the Diet is more likely, if required, to endorse the results of the negotiation, and the Japanese government's chances of convincing other countries to cooperate in reaching an agreement increase.

Analyzing a US–Japan trade negotiation

Building on these conceptualizations of American and Japanese trade policymaking, when the American and Japanese governments are negotiating between themselves concerning some trade issue, the legislature and other actors in each country may exert pressure on their negotiators, or on negotiators or other actors in the other country, which can result in the formation of transnational coalitions that have an effect on the negotiations. If additional countries other than the United States and Japan are involved in the negotiation, the legislature and other actors in the additional countries may similarly pressure their negotiators or try to influence the negotiators of or actors in the other countries, and thereby influence any transnational coalitions that form.

To fully understand a US–Japan trade negotiation it is necessary to understand not only the domestic politics related to the negotiations in each country, but also the relationship between the negotiators of the countries involved. Cultural factors, past relations, visions of future relations and conflict tendencies in the relationship between the United States and Japan can affect the negotiation. Communication and information exchange problems, information asymmetries, role allocation and incentives can also affect the negotiation.

The individual or individuals who negotiate for the United States and Japan occupy a role between the domestic and international levels. An American or Japanese negotiator must often negotiate in two directions at the same time – with certain actors in the domestic politics of his or her own country, and with the negotiator, and sometimes other actors, from the other country.

Various phases in the relationship between domestic politics and international relations in US–Japan trade negotiations

By analyzing the evolving relationship between the domestic and international factors in American and Japanese trade policymaking, various phases in the relationship between domestic politics and international relations in US–Japan trade negotiations can be discerned, although no more than a brief elaboration of these phases is possible in the paragraphs that follow.

Because of the United States' increasing trade relations with East Asian countries during the nineteenth century, the United States played a major role in ending Japan's restricted relations with the outside world during the mid-nineteenth century. Foreigners in Japan were often confined to specific settlements in designated ports from 1859 to 1899. From the 1870s to the 1890s, Japan sought to regain control over its own tariffs by revising the treaties it had signed with the foreign powers, while the foreign powers sought to revise the treaties to gain greater access to Japan. In 1899, the treaty-port system ended (J. E. Hoare, 1994, pp. 97–105) and, in 1911, the United States, along with the other foreign powers, agreed that Japan could take control of its own tariffs (Lawrence Battistini, 1953, pp. 36–43).

Between 1873 and 1960, about 20 per cent to 30 per cent of Japan's exports went to the United States (Warren Hunsberger, 1964, p. 241). Prior to the Second World War, silk was one of Japan's major exports, and oil one of the United States' major exports to Japan. During the 1920s, Japanese imports of American products, especially automobiles, lumber, building material, and oil and machinery, increased dramatically. By 1932, the United States' trade deficit with Japan had disappeared, and would not reappear until the mid-1960s. As imports of certain Japanese products to the United States increased, and Japan's colonial empire expanded, such imports to the United States became the object of some criticism. During the 1890s, the American silk industry and refiners of camphor expressed concern about Japanese imports. A surge in Japanese cotton imports from 1933 to 1936 led to calls from American producers to reduce the flow. In response, in 1936, the United States imposed a duty on imports of bleached and colored cotton and fabrics. During the late 1930s and early 1940s, both the American and Japanese governments took action to restrict the commercial activities of the other country, and to restrict trade relations between the two countries (Fairbank *et al.* 1989, pp. 718–21; Nakamura, 1997, pp. 149–50; William Neumann, 1963, pp. 212–27).

After the Second World War, Japan continued to import agricultural products, raw materials and fuel from the United States. Gary Saxonhouse and Hugh Patrick note that 'During the 1950s and into the 1960s, the Japanese economy was neither large enough nor sufficiently technologically sophisticated to be of interest to American policymakers and businessmen' (Saxonhouse and Patrick, 1976, pp. 98–9). During the 1950s, as the United States sponsored Japan's accession to the GATT, the Japanese economy recovered. Spurred on in part by the Korean War, Japanese exporters benefited from easy access to the American market. The United States imported steel, metal products, textiles and clothing from Japan (Lockwood, 1966, pp. 118–19), sparking calls among some affected American interest groups to limit imports from Japan. Under pressure from the American government, the Japanese government agreed to restrain its cotton exports to the United States between 1957 and 1961 (Hunsberger, 1964, pp. 263–327), which led to a 1961 bilateral, short-term agreement on cotton textiles, and eventually to the 1974 Multi-Fiber Agreement.

During the early 1960s, the cabinet-level Joint United States–Japan Committee for Trade and Economic Affairs was established (Frank Langdon, 1973, pp. 58–73). During the mid-1960s, Japan once more started to show a trade surplus with the United States.

During the early 1970s, as Japan's economy grew rapidly, and as the United States' global economic pre-eminence declined, an economic slowdown, inflation, unemployment and government deficits in the United States, accompanied by a steady growth in trade imbalances, caused increasing trade tensions between the United States and Japan.

Japanese exports to the United States of steel and other products, such as color televisions, attracted the attention of American producers, who pressured the US government to take action against such imports. In response, some Japanese companies invested directly in the United States (Komiya and Motoshige Itoh, 1988, pp. 197–8). The United States and Japan entered into negotiations concerning recent large increases in American imports from Japan of steel and automobiles, and concerning the difficulties American exporters were encountering in exporting beef, citrus fruits and other products to Japan (Destler and Hideo Sato, 1982a).

During the 1980s, as American hegemony declined and the challenges posed to the United States by Japan's economic growth and competitiveness increased, the two countries engaged in a series of trade negotiations, including the Market-Oriented Sector-Selective (MOSS) negotiations, which began in 1985. Japan responded to these negotiations, by making a number of trade concessions, including imposing voluntary export restraints on steel in 1984, and entering into a 1986 agreement in which Japan agreed to make efforts to increase purchases of foreign semiconductors (Thomas Pugel, 1987). Starting in 1989, the United States and Japan engaged in the Structural Impediments Initiative talks, in which the United States focused on structural impediments to trade such as Japan's savings and investment, land use, pricing, distribution, and anticompetitive and exclusionary business practices, and Japan expressed concern about American savings and investment, corporate competitiveness, promotion of exports, labor training, and education (Naka, 1996, pp. 17–24).

By the mid-1980s, when the case study examined in this book begins, the American trade deficit with Japan had increased dramatically, and a growing number of American interest groups were pressuring the Executive Branch and Congress in order to convince the United States government to take actions to increase access to the Japanese market (Merit Janow, 1994).

During the 1990s and the first few years of the twenty-first century, as the American economy continued to grow and the Japanese economy faltered, pressure remained on the American government to improve access to the Japanese market, and US–Japan trade negotiations continued, although the negotiations attracted much less attention than they had during the 1980s. At the same time, Japan showed 'a new willingness to use GATT–WTO rulemaking and rule enforcement to directly confront and challenge the United States' (Searight, 2002, p. 178).

Designing and testing hypotheses that explore the relationship between domestic politics and international relations in a US–Japan trade negotiation

Given the complexity of US–Japan relations, there are a large number of hypotheses that could be designed and tested in order to explore the relationship between domestic politics and international relations in US–Japan trade negotiations. In order to limit the scope of the inquiry, this book explores the relationship between domestic politics and international relations in modern US–Japan trade negotiations by developing and testing the following hypotheses that concern cooperation and the interaction of negotiating strategies at the international level, in order to address the question: To what extent does domestic politics affect the agreement the United States and Japan reach in a particular trade negotiation?

- First, if the American and/or the Japanese government adopt(s) a protectionist negotiating stance, particularly as a result of key elements within the state (bureaucracy) and the legislature supporting such a stance, then the government that adopts a protectionist negotiating stance is less likely to seek to cooperate concerning trade liberalization in an international trade negotiation.
- Second, the more divided the American and/or the Japanese government are concerning the negotiating stances that should be taken, the more likely the divided government will be able to alter its negotiating stance if that government chooses to cooperate in order to reach an agreement.
- Third, the more informed or key domestic groups there are in both the United States and Japan who endorse the negotiating stances of their governments, the more likely the US Congress and the Japanese Diet are, if required, to endorse the results of the negotiation, and the greater the chances of the two governments reaching an agreement.

The GATT Uruguay Round agriculture negotiations

In order to develop the contextual two-level game approach to analyzing trade policymaking and to further explore the approach's application to American trade policymaking, Japanese trade policymaking and US–Japan negotiations, it is logical to select a case study that concerns a negotiating round of the General Agreement on Tariffs and

Trade (GATT). Since 1948, when the GATT treaty came into effect, various GATT negotiating rounds, including the Kennedy Round (1963–7), the Tokyo Round (1973–9), and the Uruguay Round (1986–94), have involved an increasing number of countries whose domestic politics have affected the negotiating rounds, and whose domestic politics have been influenced by the negotiating rounds.[7]

It is logical to select a case study involving GATT negotiations concerning agriculture, because during the GATT negotiating rounds the negotiations concerning agriculture have been quite influenced by the domestic politics of the major GATT member countries, partly as a result of the 'constituency representation, strong government machinery, and effective political pressure' that characterizes agricultural trade policymaking (Gilbert Winham, 1986, p. 148).

Examining the agriculture negotiations that occurred during the GATT Uruguay Round negotiations (1986–94) is of particular relevance, because gathering sufficient domestic political support in some countries, especially Japan, for certain provisions in the Uruguay Round Agreement on Agriculture that resulted from the Uruguay Round negotiations was essential to the successful conclusion of the entire Uruguay Round negotiations. Some of the provisions of the Uruguay Round Agreement on Agriculture that were debated most actively in the domestic politics of the major GATT member countries – the provisions concerning the comprehensive tariffication of agricultural products and minimum access to agricultural markets – are the focus of Chapters 3, 4, and 5 of this book.

Chapters 3, 4, and 5 develop the contextual two-level game approach to analyzing trade policymaking by examining the GATT Uruguay Round agriculture negotiations, following the suggestion that developing a theory (or in this case, an analytic approach) by examining a case study can help 'to discern important *new* general problems, identify possible theoretical solutions, and formulate potentially generalizable relations that were not previously apparent' (Alexander George, 1979, p. 51).

Following Alexander George's (1979, p. 52) suggestion that 'a simultaneous comparison of two or more cases, if each comprises an instance of the same class of events, can be an excellent research strategy for the cumulative development of theory' (or in this case an analytic approach), Chapters 3 and 4 focus, respectively, on American and Japanese trade policymaking, during the Uruguay Round agriculture negotiations, as two instances within the same multilateral negotiation in which domestic politics played an important role.

Based on the above considerations, Chapters 3, 4, and 5 analyze American and Japanese trade policymaking, and US–Japan trade negotiations during the Uruguay Round agriculture negotiations, covering the period from the first discussions in 1982 of a new GATT negotiating round, through the September 1986 Punta del Este GATT Ministerial Meeting, during which the negotiating agenda for the Uruguay Round was established, to the 15 December 1993 conclusion of the Uruguay Round negotiations, and the late 1994 passage of implementing legislation by both the American and Japanese governments.[8] The discussion that follows highlights the following phases of the negotiations:

- The GATT Committee on Trade in Agriculture and the emerging consensus to include agriculture in the Uruguay Round negotiations (November 1982–December 1985);
- The Preparatory Committee (January–August 1986);
- The Punta del Este Declaration (September–December 1986);
- The beginning of the Uruguay Round negotiations and the presentation of the first negotiating proposals to the Uruguay Round Negotiating Group on Agriculture (January 1987–February 1988);
- The negotiations leading to the Uruguay Round Mid-Term Review Agreement, the United States' proposals concerning the comprehensive tariffication of agricultural products and Japan's elaboration of its food security stance (March 1988–June 1989);
- The tabling of national positions in the Uruguay Round Negotiating Group on Agriculture (July–December 1989);
- The negotiations culminating in Uruguay Round Negotiating Group on Agriculture Chairman de Zeeuw's July 1990 Framework Agreement (January–July 1990);
- The failure to conclude the Uruguay Round on schedule at the December 1990 Brussels Ministerial Meeting and its aftermath (August 1990–September 1991);
- The December 1991 Draft Final Act and debates concerning it (October 1991–November 1992);
- The conclusion of the Uruguay Round negotiations (December 1992–December 1993); and
- The signing of the Uruguay Round agreements and the passage of implementing legislation (January–December 1994).

3
The US and the Uruguay Round Agriculture Negotiations

In an effort to increase agricultural exports, and concerned about recent declines in the United States' agricultural trade surplus, the United States government adopted an ambitious free-trade negotiating stance in the Uruguay Round agriculture negotiations. This chapter traces the relationship between domestic politics and international relations in American trade policymaking during the Uruguay Round agriculture negotiations, focusing on the negotiations concerning market access. The chapter concludes by addressing the question: To what extent did domestic politics affect the agreement the United States reached in the Uruguay Round agriculture negotiations? and assesses the validity of the three hypotheses concerning American trade policymaking discussed in Chapter 2.

The Reagan Administration and the beginning of the Uruguay Round

The Reagan Administration advocated placing agriculture on the negotiating agenda of the Uruguay Round negotiations in an effort to halt recent declines in America's agricultural trade surplus.

During the late 1940s, when the General Agreement on Tariffs and Trade (GATT) was drafted, there was much more domestic political support for making trade in manufactured goods subject to the mutual commitments that the United States entered into in the GATT than there was for making trade in agricultural products subject to such mutual commitments. American agricultural policy was relatively market-oriented during the postwar period. However, while Article xi of the GATT called for a 'General Elimination of Quantitative Restrictions' on imports, in 1955 the United States received a GATT waiver for some of its import restrictions. These were import quotas on agricultural products that the United States had established in 1935 and later amendments to Section 22 of the Agricultural Adjustment Act of 1933. Section 22 came to protect politically influential producers of a few commodities, such as dairy products. Section 22 authorized the President to cause an investigation to be made by the US International Trade Commission which could lead the President to impose quantitative restrictions on the imports of certain articles in order to ensure that such imported articles did 'not render or tend to render ineffective or materially interfere with' any program or operation related to the US agricultural adjustment program.[1] Since 1935, using Section 22, the United States imposed import controls on twelve commodity groups, including 'wheat and wheat flour; rye; rye flour and meal; barley; oats; cotton, along with certain wastes and cotton products; certain dairy products; shelled almonds; shelled filberts; peanuts and oil; tung nuts and oil; flaxseed and linseed oil; sugars and syrups'.[2] By the mid-1980s, Section 22 import restrictions remained only on a few products, such as cotton, dairy products, peanuts, sugars and syrups.[3] The waiver that the United States received from the GATT in 1955 remedied the inconsistency between Section 22 of the Agricultural Adjustment Act of 1933 and Article XI of the GATT. Without this waiver, United States government programs implemented under Section 22 of the Agricultural Adjustment Act of 1933 could have been held to be in violation of Article XI of the GATT (US Congress, Congressional Budget Office, 1987, p. 79).

During the 1960s, America's farmers profited by exporting basic commodities as world trade in agricultural products expanded (D. Gale Johnson, 1979). But, by the late 1970s, United States' exports confronted increased competition in the world grains, citrus and rice markets (Dale Hathaway, 1983; Michael Reich *et al.*, 1986, pp. 158–60).

During March 1982, USTR William Brock informed a Senate Committee on Finance subcommittee that the United States would seek, during the upcoming November 1982 GATT Ministerial meeting, 'to bring the treatment of agriculture in the GATT more into conformity with the rules for industrial trade'.[4]

During the early 1980s, a consensus emerged that the United States should increase its agricultural exports by pressuring Japan to reduce its restrictive import policies and trade practices that limited American agricultural exports to Japan. During March 1982, Commerce Department Under Secretary for International Trade, Lionel Olmer, listed import quotas as the first of six barriers that prevented increased

American exports to Japan, and singled out Japanese import quotas that remained on twenty-two agricultural goods, noting that that these quotas had 'been in violation of Japan's GATT obligations for years'.[5] By February 1983, USTR Brock was calling for 'the total elimination' of Japan's agricultural import restrictions.[6] But some doubted whether such a strategy would lead to significant increases in American agricultural exports, or would have much of an impact on the growing American trade deficit with Japan. In February 1982, US Ambassador to Japan, Michael Mansfield, is reported to have estimated that if all the existing non-tariff barriers and tariffs on the sale of agricultural products in Japan were removed, ' "it would only amount to about $500 million over a five-year period," and would be mainly in such products as fruit puree, honey, and tomato ketchup'.[7]

The 1984 Trade Act gave the Executive Branch negotiating authority to enter into the next round of GATT negotiations. During February 1985, USTR Brock told a House Appropriations subcommittee that there was 'no possibility' that the United States would enter another GATT negotiating round without placing agriculture on the agenda.[8]

Even though American agricultural exports 'jumped from just $7.8 billion in 1971 to a record $43.8 billion in 1981', the House Committee on Agriculture, as it drafted the Food Security Act of 1985, forecast American agricultural exports to be just $32 billion for fiscal 1985. The House Committee on Agriculture noted that some of the problems confronting American agriculture in the mid-1980s resulted from factors 'largely outside the area that farmers themselves [could] control', citing 'the general world-wide recession of the early 1980s, which depressed markets for American products, the strength of the dollar in recent years, the practices of some competing nations in world markets, and continuing surpluses of some commodities'.[9]

The Food Security Act of 1985 stated that American agricultural trade policy should 'provide competitive commodities for export; support free trade; counter unfair trading practices; remove foreign policy restraints to agricultural trade; and [provide for] consideration of agricultural trade interests in the formation of national fiscal policy'. The Act created the Export Enhancement Program, which was designed to expand and develop American agricultural exports 'by countering the effects of foreign subsidies in international markets'.[10] The Act was also designed to increase American leverage and contribute to realizing an agreement concerning agricultural subsidies in the upcoming GATT negotiations.[11] The Food Security Act of 1985 created a marketing loan program for various commodities, including rice. America's market share of the world rice market declined from 23 per cent to 17 per cent between 1980 and 1985. According to the House Agriculture Committee, the short-term outlook for the American rice industry was 'bleak, dominated by large supplies, record yields, and weak demand due to the high value of the U.S. dollar and aggressive export tactics by Thailand'.[12]

The Reagan Administration considered the GATT to be the only 'action oriented' multilateral forum that the Administration could use to address the recent decline in America's agricultural exports.[13] In a September 1985 speech concerning international trade policy, President Ronald Reagan stated that his Administration would accelerate its 'efforts to launch a new GATT negotiating round with our trading partners' and hoped 'that the GATT members [would] see fit to reduce barriers for trade in agricultural products'.[14]

During April 1986, Agriculture Department Under Secretary for International Affairs and Commodity Programs, Daniel Amstutz, told the House Committee on Agriculture that improving America's agricultural export situation was 'the most critical element to the long-term revitalization' of American agriculture. Amstutz stated that the United States had three broad objectives in the upcoming GATT negotiations concerning agriculture: 'First, to improve access to foreign markets; second, to write effective GATT rules for controlling unfair trade practices such as subsidization; and, third, to harmonize the use of food, plant and animal health restrictions which impede trade.'[15] This chapter focuses on the first of these objectives – market access – and on the relationship between American domestic politics and the evolution of the United States' negotiating stance concerning access to agricultural markets.

In mid-June 1986, the 35-member National Commission on Agricultural Trade and Export Policy, which Congress had created in 1984, issued a report recommending that 'The President be authorized to enter into a new round of bilateral and multilateral negotiations to improve conditions affecting U.S. agricultural trade.'[16]

During the week before the September 1986 Punta del Este GATT Ministerial Meeting, at which the negotiating agenda for the Uruguay Round was to be agreed upon, USTR Clayton Yeutter summarized the United States' objectives concerning agriculture in the upcoming GATT negotiations as follows: 'We want to phase out import restrictions on agricultural products, treat agricultural export subsidies the same as subsidies for industrial products, and eliminate other barriers to market access in developed and developing countries.'[17]

Most of these concerns were included in the Punta del Este Declaration that resulted from the September 1986 GATT Ministerial meeting in Punta del Este, Uruguay. The Punta del Este Declaration launched what would become known as the GATT Uruguay Round negotiations. Included in the negotiating agenda set out in the Punta del Este Declaration was a commitment to 'aim to achieve greater liberalization of trade in agriculture and bring all measures affecting import access and export competition under strengthened and more operationally effective GATT rules and disciplines, taking into account the general principles governing the negotiations, by: . . . improving market access through, *inter alia*, the reduction of import barriers'.[18]

A few days before the Punta del Este Ministerial Meeting, the Rice Millers' Association, an association of twenty-seven members, including 'farmer-owned cooperatives and independently owned rice milling companies located in Arkansas, California, Louisiana, Mississippi, Texas and Florida' filed a Section 301 petition with the USTR, requesting that the US government take measures to pressure the Japanese government to remove its virtual ban on rice imports, a ban which the Association claimed had evolved since 1961.[19] The Rice Millers' Association claimed that its member firms milled 'virtually all rice produced in the United States and together with associate members account[ed] for virtually all US rice exports'.[20]

During October 1986, within the forty-five days required by Section 301 legislation, USTR Yeutter rejected the Rice Millers' Association's petition, noting that Japan had made commitments a few weeks earlier in the September 1986 GATT Uruguay Round Punta del Este Declaration to improve market access to its agricultural markets in the framework of the upcoming GATT round of multilateral trade negotiations. Yeutter reportedly conceded that '"Sensible, reasonable adjustments that would make [Japan's] rice program less trade distortive are, in our judgment, sound and rational negotiating objectives, not only for the United States but for all rice-producing nations of the world, many of which are developing nations"'. Yeutter noted that the United States 'should have a better understanding of all this by the middle of next year [1987] and warned that if Tokyo [had] not responded by then "in a forthcoming manner"', the United States would '"promptly reexamine the issue"'.[21]

During late January 1987, at the GATT in Geneva, the negotiating structure for the Uruguay Round negotiations was put into place. Fourteen separate negotiating groups were established, including one called the Uruguay Round Negotiating Group on Agriculture. The Negotiating Group on Agriculture's initial task was to identify major problems and their causes, consider 'basic principles to govern world trade in agriculture', and to submit and examine participating countries' proposals that aimed at achieving the negotiating objectives set out in the Punta del Este Declaration.[22] The work of the Negotiating Group on Agriculture was reviewed by the Group of Negotiation on Goods, which was supervised, along with the Group of Negotiations on Services, by the Trade Negotiations Committee.

During the Uruguay Round negotiations, the Negotiating Group on Agriculture met every few months through 1990 at the GATT in Geneva, and thereafter on a less regular basis. During the meetings of the Negotiating Group on Agriculture, various countries' representatives presented proposals concerning the agreement that should be reached, and the concessions their country was willing to make. While more than a hundred countries signed the Uruguay Round agreements, only about thirty presented at least one proposal to the Uruguay Round Negotiating Group on Agriculture. While all the lists of representatives attending the meetings of the Negotiating Group on Agriculture are not available publicly, in 1987, the United States' representatives to the Negotiating Group on Agriculture included members of the USTR and US Departments of Agriculture, Commerce and State.[23] While the Uruguay Round agriculture negotiations received much contemporaneous global press attention, the negotiations were confidential, and it is only recently that some of the GATT documents related to the negotiation have become available to the public.[24] While Chapter 6 focuses on the particular international context at the GATT in which the United States laid out its negotiating proposals, this chapter analyzes the evolving relationship between American domestic politics and the United States' negotiating stances concerning market access during the Uruguay Round agriculture negotiations based on what is currently publicly available, much of which results from Congressional hearings.

During March 1987, a USTR official described the United States' initial negotiating stance concerning market access in the Uruguay Round Negotiating Group on Agriculture to the House Agriculture Subcommittee on Livestock, Dairy, and Poultry, by noting that 'Our officials have publicly stated that all agricultural programs must be on the negotiating table if we are to have any chance to improve the agricultural trading system.'[25] Subcommittee members expressed concern that the United States' negotiating stance might result in the possible relaxing of Section 22 import quotas, such as those that affected dairy products. House Agriculture Subcommittee members stated that the United States' negotiating stance appeared to suggest that changes in Section 22 of the Agricultural Adjustment Act of 1933 might be necessary if the GATT waiver that the United States had received for its programs under Section 22 were modified as a result of the agriculture agreement reached during the Uruguay Round negotiations. A National Milk Producers Federation representative noted that Section 22 quotas had been applied 'to most imported dairy products', and complained that the removal of Section 22 quotas could result in 'unlimited dairy imports [that] would soon destroy the domestic dairy price support program'.[26] However, an April 1987 House Committee on Agriculture report stated that no one had 'done a complete analysis of the potential effect' on American dairy farmers of the elimination or substantial reduction of Section 22 protection.[27]

During July 1987, the Congressional Budget Office released a study that showed that falling exports and rising imports had resulted in the US agricultural trade surplus declining from a peak of $26.6 billion in 1981 to $5.0 billion by 1986. The study noted that Japan had been the 'largest single purchaser of U.S. agricultural exports', but that Japan had recently purchased 24 per cent less of American agricultural products – $5.1 billion in 1986, in comparison to $6.7 billion in 1981. The report stated that, in order to induce other countries to make concessions during the Uruguay Round agriculture negotiations, the United States would 'have to consider overhauling many of its farm support programs', including rescinding the GATT waiver which the United States had received in 1955 for its Section 22 agricultural import quotas.[28]

During the 6–7 July 1987 Uruguay Round Negotiating Group on Agriculture meeting, the United States presented its first major negotiating proposal to the Uruguay Round Negotiating Group on Agriculture. The proposal was ambitious, calling for a ten-year phase-out of import barriers, as well as a phase-out of all agricultural subsidies (including export subsidies), and an agreement concerning health and sanitary regulations. The United States proposed a two-stage process to accomplish such goals. First, 'measuring devices and an overall schedule of reductions should be agreed to for taking aggregate levels of support to zero over a 10-year period'. Second, 'specific policy changes should be identified by each country to meet its overall commitment of scheduled support reductions', and these changes should be agreed to by the member countries of the GATT.[29] The American Farm Bureau Federation, one of America's largest agricultural interest groups, is reported to have generally 'reacted favorably' to this proposal.[30]

In August 1988, with the Omnibus Trade and Competitiveness Act of 1988, Congress extended trade negotiating authority to the President until June 1991 to sign a Uruguay Round agreement with a two-year extension possible. The President was required to inform Congress by mid-February 1990 of the Executive Branch's progress in the Uruguay Round agriculture negotiations. The law also allowed the President to implement certain programs in the event that Uruguay Round implementing legislation was not enacted by 1990. For example, the act stated that if the President did not certify significant progress in the negotiations by 1990, the President could implement a marketing loan program for wheat, feed grains and soybean producers, or could instruct the Secretary of Agriculture 'to make agricultural commodities and products acquired by the Commodity Credit Corporation equaling at least $2,000,000,000 in value available during the 1990 through 1992 fiscal years to United States exporters of domestically produced agricultural commodities and products'.[31]

During September 1988, the Rice Millers' Association filed a second Section 301 petition with USTR, calling upon the United States government to take action to open up Japan's rice market. The petition stated that the United States should conclude a four-year agreement with Japan that would allow American rice farmers 'to capture 2.5 percent of the Japanese market in the first year, 5 percent in the second, 7.5 percent in the third, and 10 percent in the fourth'. It was estimated that if such a plan were implemented, American rice shipments to Japan might reach $760 million by the end of the fourth year.[32] One month later, during October 1988, USTR Yeutter once more rejected the Rice Millers' Association's petition, reportedly stating ' "I concluded two years ago that attempting to eliminate Japan's import restrictions on rice through the Uruguay Round of multilateral trade negotiations was more likely to be successful than a Section 301 case, and I still believe that to be the case." ' But Yeutter also warned that if Japan did not permit progress on this issue at the upcoming December 1988 Uruguay Round Montreal Mid-Term Review meeting, the USTR would invite the Rice Millers' Association to resubmit its Section 301 petition.[33]

During November 1988, the United States elaborated on its July 1987 proposal to the Uruguay Round Negotiating Group on Agriculture. As part of the United States' proposal concerning market access, the United States advocated the comprehensive tariffication of agricultural products, calling for Ministers to reach agreement in the Uruguay Round Mid-Term Review Agreement to 'convert all non-tariff barriers, including variable levies and barriers maintained under waivers or other exceptions, into tariffs'. The tariffs could then be reduced through ongoing negotiations. The United States also called for participants to submit by January 1990 '[s]pecific proposals for the rollback of non-tariff barriers and subsidies affecting trade'.[34]

Just before the December 1988 Montreal GATT Uruguay Round Mid-Term Review Ministerial Meeting, President Reagan told the National Chamber Foundation that he could not 'think of any other matter in the last 8 years that ha[d] achieved the same breadth of support within our government' as the GATT negotiations.[35]

During the December 1988 Montreal GATT Ministerial Meeting, be-cause the United States and the EC were unable to reach an agreement concerning the agricultural provisions of the Uruguay Round Mid-Term Review Agreement – particularly the provisions concerning agri-cultural export subsidies – the Uruguay Round Mid-Term Review Agreement was not finalized until the early months of the George H. W. Bush Administration.

The George H. W. Bush Administration and the Uruguay Round Mid-Term Review Agreement, the Draft Final Act, and the US-EC Blair House Accord

After the GATT member countries were unable to agree on the provisions of the Uruguay Round Mid-Term Review Agreement during the December 1988 GATT Montreal Ministerial Meeting, they selected an April 1989 deadline to conclude the Mid-Term Review Agreement, which meant that major decisions concerning the Uruguay Round agriculture negotiations had to be made during the first few months of the first George H. W. Bush Administration (January 1989—January 1993), as a new trade policy team, including USTR Carla Hills and Secretary of Agriculture Clayton Yeutter was being put into place.

On 3 April 1989, the International Trade Commission held a hearing to assess the potential impact of the proposed Uruguay Round Mid-Term Review Agreement on American agriculture. Most of the large American farm groups, such as the American Farm Bureau Federation and the US Feed Grains Council, supported the United States' negotiating stance and the Uruguay Round Mid-Term Review Agreement under discussion. But certain groups – for example, the National Farmers Union, the National Corn Growers Association, and the Coalition to Save the Family Farm – expressed concern about the extent to which certain provisions that had been included in drafts of the Uruguay

Round Mid-Term Review Agreement might lead to reductions in American import barriers that would affect them adversely. The National Farmers Union complained that 'the removal of Section 22 of the Agricultural Adjustment Act "could result in a flood of imports which could devastate"' certain industries.[36]

Despite such concerns, during an early April 1989 Uruguay Round Trade Negotiations Committee meeting, the United States accepted the Uruguay Round Mid-Term Review Agreement. The forty-four members of the United States' delegation to the April 1989 meeting included not only a representative from the USTR, advisers from the USTR and US Departments of Commerce, State, Agriculture and Treasury, but also seven Congressmen, including the Chairman and six members of the House Committee on Agriculture, as well as eight Congressional staff advisers, including four from the House and Senate Agriculture Committees.[37]

The agricultural provisions included in the Uruguay Round Mid-Term Review Agreement expanded on the agricultural provisions that had been included in the Punta del Este Declaration. The Mid-Term Review Agreement stated that the long-term objective was 'to provide for substantial progressive reductions in agricultural support and protection sustained over an agreed portion of time, resulting in correcting and preventing restrictions and distortions in world agricultural markets'. In realizing this long-term objective, the Mid-Term Review Agreement called for 'strengthened and more operationally effective GATT rules and disciplines, which would be equally applicable to all contracting parties'. The Agreement stated that 'the commitments to be negotiated, should encompass all measures affecting directly or indirectly import access and export competition'. Among the measures affecting import access, the Mid-Term Review Agreement listed 'quantitative and other non-tariff access restrictions, whether maintained under waivers, protocols of accession or other derogations and exceptions, and all measures not explicitly provided for in the General Agreement, and the matter of conversion of the measures listed above into tariffs;' as well as 'tariffs, including bindings'. The Mid-Term Review Agreement stated that, by October 1989, Ministers would provide specifics concerning their 'intention to reduce support and protection levels for 1990'. By December 1989, participants were to set down detailed proposals, and the goal was to agree on 'the long-term reform programme and the period of time for its implementation' by the end of 1990.[38]

During late-April 1989 Congressional hearings, an Associated Milk Producers' representative voiced concern about the potential impact on the American dairy industry of the phasing out of Section 22 quantitative restrictions on agricultural products if the United States' negotiating stance were accepted as part of the final Uruguay Round agricultural agreement.[39] American officials responded to such concerns by stating that they did not foresee any need to change the operation of US agricultural programs as a result of the United States having agreed to the short-term provisions concerning agriculture in the Uruguay Round Mid-Term Review Agreement. However, United States officials noted that the Mid-Term Review Agreement made 'it clear that all policies affecting trade [were] on the negotiating table', and that the United States had proposed to the Uruguay Round Negotiating Group on Agriculture that 'in order to provide greater transparency to import access barriers, all nontariff barriers [should] be converted to tariffs which [could] then be negotiated down in the traditional GATT manner'.[40]

During the 10–12 July 1989 meeting of the Uruguay Round Negotiating Group on Agriculture, the United States elaborated on its earlier proposals and explained the role that the comprehensive tariffication of agricultural products could play in the negotiation of the reduction of market access barriers, discussing 'the methodology and mathematical formula that could be used in order to calculate the ad valorem equivalent of a non-tariff barrier'.[41] This proposal brought protests from the National Farmers Union, who is reported to have argued that such a tariffication proposal represented '"a step toward destroying U.S. farm programs"'.[42]

During the 25–26 October 1989 meeting of the Uruguay Round Negotiating Group on Agriculture, the United States once more elaborated on its plan for the comprehensive tariffication of agricultural products. As for import access, the United States suggested that the Ministers of the GATT member countries agree that all tariffs, including those resulting from conversion of non-tariff barriers to tariffs 'would be progressively reduced to zero or low levels over a ten-year period'. The United States also proposed that the Ministers agree that '[a]ll forms of derogations from existing GATT rules would be eliminated', and that Article xi:2(c), which authorized certain quantitative restrictions in the agricultural sector, would also be eliminated. While such changes were being implemented, the United States suggested that 'a safeguard mechanism would operate to protect against import surges'.[43]

During the November 1989 Senate Finance Subcommittee on International Trade hearings, Subcommittee chairman, Senator Max Baucus (D-Montana), complained that the United States government's proposals to the Uruguay Round Negotiating Group on Agriculture seemed 'to be driven as much by an ideological, textbook commitment to free trade as by a desire to improve the position of U.S. agriculture'. Senator Baucus noted that Congress's consultations with the Executive Branch concerning the United States' negotiating stance at the Uruguay Round agriculture negotiations involved 'informing Congress of administration proposals in advance and allowing members and staff to observe negotiations'. Senator Baucus felt that the communication had 'been essentially one-way'. Senator Baucus saw 'little willingness on the part of the administration to take congressional suggestions', stating that a Uruguay Round agreement 'must demonstrate concrete benefits for American farmers, businessmen, and workers'.[44] Senator Baucus also noted that he had introduced legislation that 'would require the Administration to use Section 301 and various agricultural export programs to protect the interests of American farmers if the GATT talks [broke] down'.[45] Deputy USTR Julius Katz called such legislation 'unnecessary', stating that it was 'likely to complicate' the negotiations.[46]

Most of the agricultural industry representatives, including the Rice Millers' Association representative, who testified at the November 1989 Senate Finance Subcommittee on International Trade hearing, expressed support for the United States' negotiating stance at the Uruguay Round agriculture negotiations. The spokesman for the Rice Millers' Association expressed support for the American negotiating stance, claiming that, if Japanese import barriers to rice were removed entirely as a result of a Uruguay Round agriculture agreement, the United States 'would emerge as a major supplier, providing 1.6 to 1.8 million tons' of the '3.4 to 4.8 million tons' that various countries would supply to Japan's '10-million ton market'. This would result in a 60–65 per cent increase in US rice exports, totaling 'approximately $650 million' annually.[47]

In contrast, a spokesman for the US Sugar Cane Growers and Processors and US Sugar Beet Growers and Processors cited an August 1989 US Department of Agriculture study which stated that a Uruguay Round agriculture agreement might 'result in a 69% reduction in the unit return on sugar to U.S. producers and a 42% reduction in the quantity produced'. According to the spokesman, the study also stated that 'Wheat, rice, dairy, coarse grains, oilseeds and products, and producers of other crops would all suffer producer price losses' as a result of a Uruguay Round agriculture agreement.[48]

During March 1990 Senate Committee on Agriculture Subcommittee hearings, the Chairman of the Board of the Rice Millers' Association stated that the effect of the rice marketing loan program included in the Food Security Act of 1985 had been positive, and endorsed its continuation in what became the Food, Agriculture, Conservation and Trade Act of 1990. The Chairman of the Board of the Rice Millers' Association noted that US farm income levels were 'at record levels', US rice exports were 'near record levels', and the 'domestic utilization of rice and rice products [was] at record levels'.[49]

An April 1990 US International Trade Commission report, which had been requested by the USTR in order to assess the potential impact on American agriculture of the United States' negotiating stance concerning the comprehensive tariffication of agricultural products, included a calculation of the estimated tariff equivalents for existing United States' agricultural import quotas. The report examined US government programs and policies that concerned sugar, meat, peanuts, cotton and dairy products, including quotas that had been set up pursuant to Section 22 of the Agricultural Adjustment Act of 1933. The report found *ad valorem* tariff equivalent estimates which ranged from for example 30 per cent to 50 per cent in 1986, to 50 per cent to 90 per cent in 1988 for peanuts, from 160.6 per cent in 1986 to 64.5 per cent in 1988 for dry whole milk, from 142.5 per cent in 1986 to 5.7 per cent in 1988 for nonfat dry milk, from 190.2 per cent in 1986 to 95.9 per cent in 1988 for butter, and from 123.5 per cent in 1986 to 47.3 per cent in 1988 for cheddar cheese.[50]

During July 1990, Aart de Zeeuw, the Chairman of the Uruguay Round Negotiating Group on Agriculture, released a draft Uruguay Round agreement for agriculture which included several concepts that the United States had set out in its negotiating proposals to the Uruguay Round Negotiating Group on Agriculture. De Zeeuw's Framework Agreement, while it received a great deal of global press attention, was not derestricted until 1997. The draft agreement called for country lists to be submitted not later than 1 October 1990. These lists were supposed to be built on a number of parameters, including 'the conversion of all border measures other than normal custom duties into tariff equivalents, irrespective of the level of existing tariffs'. The draft agreement also called for 'in the case of absence of significant imports, establishment of a minimum level of access from 1991–1992'. This minimum access level was to be based on 'tariff quotas at low or zero rate and representing at least [x] per cent of current domestic consumption of the product concerned'.[51] As a USTR official explained it to a House Committee on Agriculture Subcommittee several months later, de Zeeuw's Framework Agreement called for the reduction of tariffs on agricultural products, but also for minimum access levels to be increased during a transition period, the length of which still needed to be negotiated.[52]

As a result of the inclusion of such provisions, it became more likely that similar provisions would be included in the final Uruguay Round agreement concerning agriculture that might result from the Uruguay Round negotiations. This increased the likelihood that the United States would have to adjust some of its agricultural programs in order to implement any resulting agreements.

Some of the United States legislation that concerned agricultural import restrictions was already subject to international pressure. In September 1990, President Bush authorized conversion of the United States' sugar import quota program, which had been created by proclamation in 1974,[53] into a tariff-based quota system after a June 1989 GATT dispute panel brought by Australia found that US quantitative import restrictions on sugar were inconsistent with US obligations under the GATT.[54] The American Sugar Alliance in October 1990 officially withdrew its support for the US negotiating stance at the Uruguay Round agriculture negotiations.[55]

During October 1990, the United States presented its offer of agricultural concessions to the Uruguay Round Negotiating Group on Agriculture calling for the Ministers to agree to '75 percent reductions each in internal supports and market access barriers'. The United States also proposed, in accordance with de Zeeuw's Framework Agreement, that, along with agreeing to comprehensive tariffication of agricultural products, countries should agree to a certain level of minimum access to their agricultural markets. The United States suggested that countries currently having import restrictions on agricultural products 'should start to import at least 3 percent of their annual domestic consumption of any import-restricted item with no or few tariffs; [and] tariffs on the remaining 97 percent should be lowered in 10 years to 50 percent'. Furthermore, this '3 percent free-import ceiling should be gradually raised to 5.25 percent in 10 years, after which all imports of the item should be conducted with tariffs under 50 percent'.[56]

The Omnibus Budget Reconciliation Act of 1990 that Congress passed during November 1990 included a safety net for American agriculture in the event that the Executive Branch did not conclude by 30

June 1992 an agricultural agreement as part of the Uruguay Round negotiations. In the event that a Uruguay Round agreement was not reached by 30 June 1992, the Act authorized the Secretary of Agriculture to lift some restraints on farm spending or increase spending on export subsidies by $1 billion beginning 1 October 1993. If no Uruguay Round agreement was reached by 30 June 1993, the law said that 'agricultural price support and other programs and export promotion levels' should be reconsidered.[57]

Just before the December 1990 Brussels GATT Ministerial Meeting, during which the Uruguay Round negotiations were scheduled to be concluded, the president of the American Farm Bureau Federation emphasized that no agriculture agreement was 'better than a bad one'.[58] As will be discussed in greater detail in Chapter 5, one of the major points of disagreement at the December 1990 Brussels GATT Ministerial Meeting was agriculture.

After the failure to conclude the Uruguay Round negotiations at the December 1990 Ministerial Meeting, US Secretary of Agriculture Yeutter resigned. The president of the American Farm Bureau Federation, among others, felt that Yeutter's resigning in order to serve as Republican Party chairman 'would not help U.S. credibility in the Uruguay Round'.[59]

During February 1991 House Committee on Agriculture Subcommittee hearings, a USTR official discussed the domestic implications of the comprehensive tariffication and minimum access provisions that the United States had proposed in October 1990 to the Uruguay Round Negotiating Group on Agriculture. The official explained that the United States' proposal for comprehensive tariffication of agricultural products, if included in the final Uruguay Round agriculture agreement, would require that non-tariff trade barriers, such as the quotas the United States maintained under Section 22 of the Agricultural Adjustment Act of 1933, would have to be converted into tariffs that would provide the same level of protection as the current Section 22 quotas did currently. The official also stated that because the United States had proposed that minimum access levels for agricultural products initially be established 'at 3 percent of domestic consumption' for products such as peanuts that were subject to Section 22 quotas, were the United States' proposal accepted, the United States would be required to import at least 3 per cent of domestic consumption of such products.[60] During the same hearing, an official from the US General Accounting Office noted that American 'sugar, dairy products, peanuts, and cotton' commodity groups were among the 'most fearful of the impact of trade liberalization' that might be brought about by a Uruguay Round agriculture agreement, because they had been protected significantly by import quotas.[61]

A few months later, as Congress debated extending for two years the President's fast-track negotiating authority in accordance with the provisions set out in the Omnibus Trade and Competitiveness Act of 1988, USTR Hills further explained the United States' minimum access proposal to the House Committee on Agriculture. USTR Hills stated, 'We started with a focus of 3 percent, really in balance with some of our markets which have less than 3 percent access, and felt that would start the reform process. Of course, that is the initial minimal access, which grows under our proposal, which has not been bought, by 0.7 percent, so over a 5-year period of time it roughly doubles.'[62] The American Farm Bureau Federation was in favor of extending negotiating authority to the Bush Administration, but the Federation noted that it would not accept any Uruguay Round agreement in which American farmers were 'asked to give up more than they receive from other countries'. The Federation noted that, for market access, 'A good Uruguay Round agreement would (1) establish clear GATT rules and disciplines on the use of variable import levies, and (2) lower or remove other forms of barriers to U.S. farm exports.' The Federation acknowledged that 'The Uruguay Round held out the best possibility for opening the Japanese rice market without resorting to bilateral dispute settlement.'[63] During the same hearings, the Rice Millers' Association focused on access to the Japanese rice market, noting that 'the Government of Japan bans virtually all rice imports, denying U.S. farmers and millers access to a 10 million ton market', and quoting the USTR as estimating that 'this ban costs the industry in excess of $650 million annually'.[64]

Meanwhile, representatives of other agricultural sectors opposed the extension of fast-track negotiating authority, expressing concern about the domestic implications of the United States' Uruguay Round negotiating proposals. The American Agriculture Movement stated that were Section 22 import restrictions removed as a result of commitments made during the Uruguay Round agriculture negotiations, 'We would lose several of our most successful farm programs', including '[q]uotas on peanuts, dairy, sugar, tobacco, and beef, among others'.[65] The Farmers Union Milk Marketing Cooperative described the potential loss of the Section 22 quota as 'an extremely serious threat to the nation's dairy industry'.[66] The National Farmers Union noted that a more positive US negotiating stance concerning market access would 'recognize the right of sovereign countries to develop their own domestic food policies, using tools such as our Section 22'.[67]

During May 1991 Senate Agriculture Committee hearings, USTR Hills stated that the government would 'make no change in section 22 unless the other countries ma[d]e substantial changes in their variable levies, their bans on imports, and their various restrictions'.[68] In response to a question concerning the possible negative effect on the American sugar industry of requiring minimum access of 3 per cent to 5 per cent, US officials indicated that a 3 per cent or 5 per cent minimum access commitment would not affect sugar 'because the import to consumption ratio for sugar [was] already much higher than 5 percent'.[69]

During May 1991, Congress voted to extend until 2 March 1993 the administration's fast-track negotiating authority. The Senate, by a vote of 59–36, rejected a resolution that would have denied the extension. The House, by a vote of 231–192, rejected a similar resolution.[70]

During the early December 1991 House Agriculture Committee hearings, as GATT Director-General Arthur Dunkel prepared the December 1991 Draft Final Act of the Uruguay Round agreements, Deputy USTR Katz outlined the agreement that had been reached in Geneva concerning the comprehensive tariffication of agricultural products. According to Katz, agreement had been reached on the need to convert non-tariff measures into tariffs, 'as the primary instrument of reform of import policies'. Katz noted that agreement had also been reached that 'Such tariff equivalents would be subject to formula reductions.' Katz stated that, furthermore, the United States had reached general agreement with the EC and the Cairns Group that 'there should be the opportunity for [market] access of at least 3 percent of domestic consumption initially, and this minimum access commitment should grow in subsequent years'. Also, there was agreement 'to ensure minimum access opportunities where no access ha[d] existed in the past'.[71]

During the same hearings, the National Family Farm Coalition complained that Section 22 was on the 'chopping block' because of the Japanese rice issue.[72] A National Peanut Growers Group representative stated that 'Producers, the USDA, the [U]STR and university analysts all agree the proposed GATT agreement will result in peanuts being the big loser.'[73]

The December 1991 Draft Final Act of the Uruguay Round agreement released by GATT Director-General Arthur Dunkel included many elements of the United States' proposals to the Uruguay Round

Negotiating Group on Agriculture. The agricultural provisions of the Draft Final Act stated that tariff equivalents, or tariffication should occur for 'border measures other than ordinary customs duties'. Tariffication would affect 'quantitative import restrictions, variable import levies, minimum import prices, discretionary import licensing, non-tariff measures maintained through state trading enterprises, [and] voluntary export restraints'.[74] The Draft Final Act stated that 'Ordinary customs duties, including those resulting from tariffication, shall be reduced, from the year 1993 to the year 1999, on a simple average basis by 36 per cent with a minimum rate of reduction of 15 per cent for each tariff line.' The Draft Final Act also stated that 'Where there are no significant imports, minimum access opportunities shall be established. [Such minimum access opportunities] shall represent in the first year of the [six- to seven-year] implementation period not less than 3 percent of corresponding domestic consumption . . . and shall be expanded to reach 5 percent of that base figure by the end of the implementation period.'[75]

Shortly after the December 1991 Draft Final Act was released, a US Department of Agriculture official informed the House Committee on Agriculture that if the United States agreed to the minimum access provisions included in the Draft Final Act, the United States would have to remove all its import restrictions that prevented less than 3 per cent of 1986–8 domestic consumption of agricultural commodities from being imported. The official noted that, in order to comply with this provision, Japan would have to allow imports of rice, and South Korea and other countries would have initially to allow at least 3 per cent of certain agricultural commodities to be imported, and then gradually allow imports of 5 per cent by the end of an agreed-upon period.[76]

The president of the American Farm Bureau Federation expressed support for the Draft Final Act, and noted that concerning minimum access, 'What pops into our mind always first is rice into Japan.' However, the Federation's president expressed concern about what the Draft Final Act 'would mean to our import-sensitive commodities', and asked 'How will the four commodities subject to the section 22 import quotas be affected and what types of income protection programs, if necessary, may be made available to them?'[77]

During late March 1992, Deputy USTR Katz informed the House Agriculture Committee that a few weeks earlier the United States had submitted 'comprehensive and complete' schedules for the United States to the GATT Secretariat concerning agricultural market access, internal support and export subsidies, following the precise format set out in the Draft Final Act. Katz stated that the United States had done this 'after consulting with members of our 11 agricultural advisory committees and congressional staff from the House and Senate Agriculture and Trade Committees'.[78]

As the negotiations between the United States and the EC over the agricultural provisions in the Draft Final Act dragged on through the summer and fall of 1992, the president of the Rice Millers' Association threatened once more to file a Section 301 petition with the USTR in order to have the United States government pressure Japan to open up its rice market. The Rice Millers' Association was concerned that if the Uruguay Round negotiations were not concluded successfully, a commitment to minimum access to agricultural markets would not be agreed upon, and American rice producers would continue to be denied access to Japan's rice market.[79]

After the November 1992 Presidential elections, the Bush Administration reached the Blair House Accord with the European Community. The Blair House Accord included 'jointly-presented proposals to amend elements of the Draft Final Act with respect to agriculture'.[80] These proposals mainly concerned reductions in domestic support and export subsidies, and the oilseeds dispute. The Blair House Accord represented a major step forward in concluding not only the Uruguay Round agriculture negotiations, but also the Uruguay Round negotiations.

The Clinton Administration and the conclusion of the negotiations, signing of the agreements, and passage of implementing legislation

Before assuming the position of USTR in the incoming Clinton Administration, during January 1993 nomination hearings, Mickey Kantor stated, concerning the Uruguay Round agriculture negotiations, that 'No deal is better than a bad deal.' But he noted that if the GATT contracting parties were 'unable to reach an agreement on agriculture', there would be 'no GATT deal'. Kantor wanted to focus on some of the provisions, such as market access, 'that were not addressed in the Blair House Accords'.[81]

During late February 1993, incoming Agriculture Secretary Mike Espy reportedly cautioned that it 'would be "really extreme"' for the United States to commence a Section 301 action against Japan concerning access to Japan's rice market. Espy pledged to resolve the Japanese rice market issue through the Uruguay Round negotiations, although he acknowledged that some bilateral meetings might be necessary.[82]

During early March 1993, members of the Rice Millers' Association met with USTR Kantor and David Graves, president of the Rice Millers' Association, is reported to have urged the Clinton Administration to explore '"whether the administration would be interested in working on a bilateral solution"' to ending Japan's virtual ban on foreign rice imports.[83]

During March 1993, USTR Kantor told the House Committee on Agriculture, 'If we obtain good results on market access – cutting tariffs, breaking down non-tariff barriers – the [Uruguay] Round will offer significant potential benefits for the American farm community.' Kantor noted that the US Department of Agriculture had estimated 'that a successful Uruguay Round agreement would expand U.S. agricultural exports by $6 to $8 billion annually after 5 years, and add $1 to $2 billion to farm income'. Kantor also stated that 'The three-year deadlock between the rest of the world and the EC over agriculture stalemated the [Uruguay] Round and gave other nations, most notably Japan, the ability to avoid contributing meaningfully to the successful completion of the talks.'[84] Kantor believed that 'if we can open markets and provide market access tariffication in agriculture and other areas that we will prosper, lead global growth, and our workers will benefit from it'.[85]

A few months later, during May 1993 Senate Committee on Finance hearings, USTR Kantor stated, 'What we have attempted to do is, to put market access first.' Kantor noted, 'What had been the case in the past 6 years is the Europeans and others had held market access to the end of the process, and the problem of the U.S. negotiators was that the most important aspect for us, market access, whether it be agriculture, industrial products or services, was being held to the end where we did not know what the end game was.'[86] Kantor also stated that agriculture was just 'a linchpin' rather than 'the linchpin', as USTR Hills had stated, for the entire Uruguay Round agreement.[87]

During Spring 1993, the Clinton Administration asked Congress for, and received in June 1993, an additional extension of fast-track negotiating authority. The extension allowed the President to negotiate a Uruguay Round agreement up to 16 April 1994, under the condition that the President was to notify Congress not later than 15 December 1993 of his intention to enter into a new GATT agreement,[88] which meant that the negotiations should be concluded by 15 December 1993 at the latest.[89]

During an early-July 1993 Toronto quadrilateral meeting, the United States, the EC, Japan and Canada reached a general agreement concerning market access.[90] During the same month, USTR Kantor admitted to a House Ways and Means Subcommittee that, in agriculture, 'we need to make tremendous progress in the area of tariffication and the areas of both current and minimum access'.[91] Kantor emphasized that, for the United States, agriculture 'remains a central component of a final Uruguay Round package'. Kantor noted that at the July 1993 Tokyo summit the G-7 countries had agreed 'that work must proceed in Geneva to finalize country schedules and the details of market access commitments'. Kantor stated that 'With the Blair House agreement behind us, market access is the major issue for Geneva negotiations. The United States is at common cause with our trading partners in Cairns who insist that the access achieved at the conclusion of the Round must be greater than the access that currently is being provided', and that this principle would continue to guide 'our negotiations'.[92]

As the Uruguay Round negotiations neared their conclusion, during the November 1993 House Ways and Means Subcommittee hearings, an American Corn Growers Association representative, speaking on behalf of the National Farmers Union and the National Family Farm Coalition, stated that 'One of the problems that we have contained both in the Dunkel [Draft Final Act] and Blair House accords is the removing of section 22, and the shift from controlling imports to tariffication.' The representative stated that, as a result of such agreements, 'we will be replacing effective laws with ineffective tariffs' which are subject to 'huge shifts in the values of currencies'. The representative stated 'We just do not believe that tariffication will work on a regular and consistent basis to protect our agriculture producers', and mentioned that sixty US Senators had written 'to the previous Administration informing them that they would not accept any attempt by GATT to use tariffication to weaken Section 22'.[93]

Despite such concerns, on 7 December 1993, a senior trade official reportedly stated that the US was prepared '"to go forward to complete the round on the basis of the understanding [in agriculture] that [had been reached]"', but '"not to do anything more"'.[94]

On 15 December 1993, the Uruguay Round negotiations were concluded, and delegates to the GATT approved a 400-page version of the Final Act Embodying the Results of the Uruguay Round. The Agreement on Agriculture that was included in the Final Act Embodying the Results of the Uruguay Round was quite similar to the 'Text on Agriculture' of the Draft Final Act,[95] and included many of the provisions that the United States had called for in its proposals to the Uruguay Round

Negotiating Group on Agriculture. The Agreement on Agriculture required the comprehensive tariffication of agricultural products, according to which non-tariff border measures were to be replaced by tariffs that provided substantially the same level of protection as existing non-tariff border measures. 'Tariffs resulting from this "tariffication" process, as well as other tariffs on agricultural products, [were] to be reduced by an average 36 per cent in the case of developed countries and 24 per cent in the case of developing countries'. The Agreement on Agriculture 'also provide[d] for the maintenance of current access opportunities' and for 'the establishment of minimum access tariff quotas (at reduced-tariff rates) where current access [was] less than 3 per cent of domestic consumption. These minimum access tariff quotas [were] to be expanded to 5 per cent over the implementation period.'[96]

Some of the domestic impact of these provisions, as reportedly summarized by Agriculture Secretary Espy at the close of the negotiations, was that the United States became 'legally bound to import 1.25 million tons of sugar', had to open its cotton market to imports, and had to import up to 5 per cent of its consumption of peanuts as it introduced the tariffication of peanuts during a six-year implementation period (the first time the United States had imported peanuts).[97]

On 15 December 1993, President William Clinton notified the Congress of his intent to enter into the agreements that resulted from the Uruguay Round negotiations, and of his intent to submit the agreements and implementing legislation for legislative approval. The President noted that the Agreement on Agriculture would 'achieve, as Congress directed, more open and fair conditions of trade in agricultural commodities by establishing specific commitments to reduce foreign export subsidies, tariffs and non-tariff barriers and internal supports'.[98] Most farm groups were generally supportive of the agreements reached.[99]

On 15 April 1994, the United States signed the Uruguay Round Agreements, including the Agreement on Agriculture, along with 100 other nations, in Marrakesh, Morocco.[100] As a result of having signed the Uruguay Round Agreements, the United States was obligated to implement the comprehensive tariffication of agricultural products and to establish a minimum level of access for agricultural products. During March 1994 hearings, Agriculture Secretary Espy informed the House Agriculture Committee that, as a result of such provisions, the United States would 'no longer be able to use the section 22 mechanism', and the tariffication of peanuts would be required.[101]

During 1994, as the American government prepared and debated the Uruguay Round Agreements Act, agricultural groups generally expressed support for many of the provisions in the Uruguay Round Agreement on Agriculture. During April 1994, a Rice Millers' Association representative told the House Agriculture Committee that 'While the [Uruguay Round] agreement [fell] short of the original U.S. objectives, the industry fully support[ed] the agricultural provisions because of the progress they [made] in providing greater access to rice markets of GATT member countries.'[102] But other groups, such as the National Farmers Union continued to complain about the minimum access provisions. An American Corn Growers Association representative stated that, for products such as 'dairy, peanuts, soybeans', of which the United States imported 'less than 5 percent' of domestic consumption, the Uruguay Round Agreement on Agriculture would mean that 'we will be competing for our own domestic market in addition to trying to go ahead and penetrate into the world market'.[103]

Despite such concerns, the provisions concerning comprehensive tariffication and minimum access included in the implementing legislation that President Clinton transmitted to the Congress in late September 1994 were not believed to have many adverse effects on American agriculture. As a result of the Uruguay Round Agreements Act[104] implementing legislation, American non-tariff import barriers became subject to comprehensive tariffication and minimum access commitments. In order to carry out comprehensive tariffication, the Statement of Administrative Action of the Uruguay Round Agreements Act stated that 'all quantitative import restrictions maintained under Section 22 of the Agricultural Adjustment Act' would be converted to tariff-rate quotas.[105] The Statement of Administrative Action emphasized that as a result of the comprehensive tariffication and minimum access provisions included in the Uruguay Round Agreement on Agriculture, the United States would benefit from other countries being required to convert quantitative import restrictions to tariff-rate quotas, and being required to adjust to specific guaranteed access levels for agricultural products. US agricultural exporters might benefit from the removal of 'the European Union's variable levy on poultry (access of 29,000 tons), the Philippines' import ban on pork (access of 54,000 tons), [South] Korea's restrictive licensing of corn (access of 6,102,100 tons), and Poland's state trading enterprise's limitations on prunes (access of 1,000 tons)'. Also, the Statement of Administrative Action noted that Japan and South Korea had agreed to provide access to their rice markets in a special treatment clause included in

Annex 5 (which concerns Article 4(2)) of the Agreement on Agriculture.[106] During April 1994 hearings, Agriculture Secretary Espy noted that American rice sales to Japan had almost reached '400,000 metric tons' and that 'These exports to that market in Tokyo will be made permanent as a part of the GATT.'[107]

Because there were perceived to be a large number of economic benefits that might result from such provisions,[108] the comprehensive tariffication and minimum access provisions were not very controversial and did not attract much attention during the Congressional debates concerning the passage of the Uruguay Round Agreements Act. In late November 1994, the House passed the Uruguay Round Agreements Act by a 288–146 bipartisan vote, and the Senate passed the Uruguay Round Agreements Act on 1 December 1994 by a 76–24 vote.

Domestic politics and international relations in American trade policymaking during the negotiations

Based on the preceding discussion of American trade policymaking during the Uruguay Round agriculture negotiations, the validity of the three hypotheses described in Chapter 2 concerning the extent to which domestic politics affects the agreement that the United States reaches in a particular trade negotiation can be assessed.

First, since key elements within the Executive Branch and Congress supported the United States in outlining an ambitious free-trade market access negotiating stance which came to include calls for an agreement on the comprehensive tariffication of agricultural products and minimum access to agricultural markets, the United States' negotiating stance during the Uruguay Round agriculture negotiations met with relatively little serious protectionist opposition within either the Executive Branch or Congress.

The major objective of the coalition within the Executive Branch and Congress that favored incorporating agriculture in the negotiating agenda of the Uruguay Round negotiations was to reduce the threat to American agricultural exports of European agricultural subsidies, through the negotiation of a mutual reduction of export subsidies. Elements of this coalition had earlier attempted to negotiate similar multilateral agreements concerning agriculture during the previous GATT Kennedy and Tokyo Rounds of negotiations. Another widely shared goal of this coalition was to increase American agricultural exports by improving access to agricultural markets. As the negotiations progressed, American negotiators elaborated how import restrictions on agricultural products could be phased out, suggesting a series of proposals detailing how the comprehensive tariffication of agricultural products and minimum access to agricultural markets could be achieved.

While the extent of agricultural reform that the United States advocated became increasingly specific in the United States' successive negotiating proposals, the comprehensiveness of the reform espoused by the United States was not substantially altered throughout the negotiations. The United States government's overall negotiating stance was largely determined within the Executive Branch during the Reagan Administration, and implemented during the Bush and Clinton Administrations. Within the Executive Branch, the Office of the US Trade Representative and the Department of Agriculture played the largest roles in developing and defending the negotiating stance that the United States adopted in the Uruguay Round agriculture negotiations. While the USTR was concerned with the entire Uruguay Round negotiations, the Department of Agriculture was particularly concerned with increasing agricultural exports, and with meeting the demands of various agricultural groups that might be affected by the ongoing negotiations. Tensions between the USTR, Department of Agriculture and other departments concerning the stances that the United States should take concerning comprehensive tariffication and minimum access were not very visible to the general public, particularly in comparison to the widespread public debate that occurred concerning the same subjects in Japan. This will be discussed in Chapter 5.

As for the validity of the second hypothesis, while the House and Senate Committees on Agriculture listened to the concerns of agricultural groups that might be affected adversely by the American government's proposals concerning comprehensive tariffication and minimum access, the general lack of division within the Executive Branch and the Congress concerning the negotiating stances that the United States adopted at the Uruguay Round Negotiating Group on Agriculture reduced Congress's influence on the negotiated agreement, and reduced the likelihood of the American government altering its negotiating stance in order to reach an agreement.

To the extent that Congress could influence the Executive Branch concerning the negotiations, the United States' negotiating stance became closer to Congress's preferences. One way in which Congress exerted influence was by monitoring the progress of the negotiations closely by holding numerous hearings, at which officials from the Executive Branch and interest group representatives testified. These hearings made American negotiators more aware of American agricultural groups' interests, needs, resources and capabilities.

Congress also influenced the Executive Branch by threatening to deny the extension of fast-track negotiating authority. The negotiations were conducted under the fast-track authority granted to the Executive Branch by the US Congress in the 1984 Trade Act, and were continued under the negotiating authority granted to the Executive Branch in the Omnibus Trade and Competitiveness Act of 1988. In 1991, Congress granted the Executive Branch the two-year extension that was provided for in the Omnibus Trade and Competitiveness Act of 1988. In early 1993, the Clinton Administration successfully requested that fast-track negotiating authority be extended to allow the negotiations to continue until December 1993, and allow for a signing of the Uruguay Round agreements by April 1994. During the 1991 hearings concerning the extension of negotiating authority, members of Congress sometimes echoed the concerns of various agricultural groups about the potential negative impact on certain agricultural producers of the comprehensive tariffication and minimum access provisions under discussion in Geneva.

Congress could also influence the impact of the negotiated agreement on American agriculture by arguing for changes in the Uruguay Round Agreements Act implementing legislation. However, under the fast-track negotiating authority, while Congress could alter the implementing legislation of the Uruguay Round agreements, Congress could not alter the Uruguay Round Agreement on Agriculture itself.

As for the validity of the third hypothesis, since the demands of interest groups such as the Rice Millers' Association, as well as the demands of many other agricultural interest groups, as represented by larger groups such as the American Farm Bureau Federation, were mostly satisfied by the comprehensive tariffication and minimum access provisions included in the United States' negotiating proposals and in the successive drafts of the Agreement on Agriculture prepared during the Uruguay Round negotiations, Congress was more likely to endorse the results of the Uruguay Round agriculture negotiations, and the chances of the United States convincing other countries to cooperate in reaching an agreement increased.

Agricultural interest groups that sought greater access to foreign markets supported the American government's efforts to negotiate comprehensive tariffication and minimum access commitments in the Uruguay Round agriculture negotiations. The Rice Millers' Association, by filing, or publicly threatening to file, a Section 301 petition, pressured the Executive Branch repeatedly in the hope that the US government would force Japan to open up Japan's closed rice market. In contrast, other farm interests, such as peanut growers and dairy producers, that might be affected adversely by a mutually agreed upon reduction in import restrictions, opposed the United States' negotiating stance because such measures might result in increased import competition.

The United States' negotiating stance, while it was mostly developed within the Executive Branch, was formed in response to ongoing pressure from the US Congress and American agricultural interest groups. The broad domestic consensus that supported the United States' negotiating stance allowed American negotiators to sustain approximately the same stance concerning market access throughout the entire negotiations, and increased the United States' chances of convincing other countries to cooperate in reaching an agreement. While certain interest groups that might suffer adverse effects, such as peanut growers, frequently raised objections to the United States government's negotiating stances concerning comprehensive tariffication and minimum access, these groups were unable to generate sufficient support within Congress to make Congress put pressure on the Executive Branch to alter the United States government's negotiating stance.

In the end, the changes in US legislation resulting from the comprehensive tariffication and minimum access provisions that the United States agreed to in the Uruguay Round Agreement on Agriculture had a negative impact on only a few American agricultural producers. Given the open nature of the United States market, the concessions involved were not that controversial within the United States. While some American agricultural interest groups were affected adversely by such changes in US legislation, other groups wished that the Uruguay Round had achieved more in terms of market access (United States Agriculture Technical Advisory Committee, 1994). However, the concessions made were widely discussed, had already been anticipated, and did not meet with much protest during the Congressional debates concerning the passage of the Uruguay Round Agreements Act.[109]

4
Japan and the Uruguay Round Agriculture Negotiations

While the debates in the United States concerning the possible impact on American agriculture of the market access provisions under discussion in the Uruguay Round agriculture negotiations were of relatively minor importance in the United States in relation to the entire Uruguay Round negotiations, similar discussions were of much greater importance to Japanese politics and to Japanese trade policymaking, and attracted the attention of the Japanese public and major Japanese newspapers for several years. This chapter examines the relationship between domestic politics and international relations in Japanese trade policymaking throughout the Uruguay Round agriculture negotiations, focusing on the negotiations concerning market access. The chapter concludes by addressing the question: To what extent did domestic politics affect the agreement reached by Japan in the Uruguay Round agriculture negotiations? This is done by assessing the validity of the three hypotheses concerning Japanese trade policymaking discussed in Chapter 2.

The Nakasone Administration and the beginning of the Uruguay Round

Agricultural reform emerged as an important issue in Japan in the late 1970s and early 1980s, and became one of the major administrative reforms espoused by Yasuhiro Nakasone, who was Prime Minister from November 1982 to November 1987. Government expenditures on rice, along with government expenditures on the national railway and national health insurance, had come to be considered as one of the three main sources of Japan's budget deficit (Frank Schwartz, 1998, p. 217).

Before the Pacific War that was part of the Second World War, Japan imported rice from Korea, Taiwan and other countries. As a result of various food shortages during Japan's Pacific War, the Japanese government tried to control through the implementation of the 1942 Food Control Act all the marketing, including the pricing, of staple foods such as rice, in order to ensure an adequate food supply for Japan's populace. After the Second World War, during the American Occupation of Japan, land reform was implemented in order to reduce hunger, create employment and democratize rural Japan (Toshihiko Kawagoe, 1993).

As the Japanese economy grew rapidly between the 1960s and the 1980s, Japan became increasingly less self-sufficient in agriculture and the world's net largest food importer. As the country came to rely on imported agricultural products, the Japanese government attempted to preserve Japan's self-sufficiency in certain consumer commodities, such as rice, as well as to maintain the income of farmers relative to urban workers through the 1961 Basic Agricultural Law. One of the foundations of the one-party dominant political system that emerged under the Liberal Democratic Party (LDP) was a rice policy that maintained government control over the producer and consumer prices of rice, and virtually forbade rice imports. The producer price of rice was sometimes adjusted upward just prior to Diet elections. As a result of efforts to support the income of farm households, the government purchase price for rice rose to several times the world price (Hideo Sato and Gunther Schmitt, 1993, pp. 255–73).

From the late 1970s to the mid-1980s, efforts were made to create more of a full-time system of commercial farming that would increase agricultural productivity and more closely match Japan's domestic agricultural production to consumption. Partly as a result of increasing pressure from various countries, including the United States, for access to Japan's agricultural markets, gradual structural adjustments, such as the liberalization of import restrictions on beef and oranges, were made (Reich *et al.*, 1986, pp. 170–87). During the GATT Tokyo Round negotiations (1973–9), increases in Japan's import quotas on beef, oranges and citrus items were negotiated (Hideo Sato and Timothy Curran, 1982).

After the December 1983 election, Japan's most important postwar political party, the LDP, held only 250 seats in the Lower House of the Diet, six less than an absolute majority. One major Japanese newspaper noted that the LDP had encountered difficulties during the election campaign because Japan's agricultural organizations feared the LDP would give in to the United States after the election in the ongoing negotiations concerning the liberalization of the beef and orange markets. The newspaper noted that there was a feeling among Japan's farmers that the LDP must be made to suffer a loss so that the party would firmly oppose agricultural concessions to the United States. During the election campaign, the opposition Japan Socialist Party and the Japan Communist Party tried to win rural votes by opposing the agricultural liberalization under discussion.[1] Despite protests from agricultural interest groups, the Japanese government reached an agreement with the United States during 1984 that included a four-year schedule for the expansion of import quotas on beef and citrus products (Amelia Porges, 1994, p. 237).

While the Japanese government strove to make Japan self-sufficient in rice by controlling the producer and consumer prices of rice, and virtually banning rice imports, poor harvests occasionally forced the Japanese government temporarily to relax its virtual ban on rice imports. During the summer of 1984, 150,000 tons of rice were imported from South Korea in order to compensate for a temporary rice shortage. Pressured by agricultural constituents who supported the Japanese government's rice policy, both Houses of the Diet adopted resolutions calling for the government to maintain its policy of self-sufficiency in rice.[2]

During June 1985, the Nakasone Administration announced an Action Program to increase imports. This included tariff reductions on some agricultural products. The Nakasone Administration stated that in the upcoming round of GATT negotiations, which the Administration favored initiating, the Japanese government would promote tariff negotiations, taking into consideration the special characteristics of agriculture, as well as the need to reduce tariffs and other factors.[3]

The April 1986 Maekawa Commission Report of the Advisory Group on Economic Structural Adjustment for International Harmony set out medium- and long-term policy measures 'concerning Japan's economic and social structure and management in response to the recent environmental changes surrounding Japan in the international economic situation'. The report proposed promoting 'agricultural policies befitting an "age of internationalization"'. The report stated that 'priorities should be given to the policies focused on fostering core farmers for the future, and price policies should be reviewed and rationalized toward greater use of market mechanisms and active promotion of structural policies'. The report noted that 'With the exception of basic farm products, efforts should be made toward a steady increase in imports of products'. The report also stated that 'With regard to products subject to quantitative import restrictions, efforts should be made to improve market access' in order to make 'the Japanese market more open while taking account of developments in the relevant consultations and negotiations including the GATT New Round'.[4] Such agricultural policy measures were particularly popular in business circles. As one scholar noted a few years later, such measures were intended to scale down 'the deficit-ridden food control system', making it possible 'to lower prices of farm produce at a single stroke, provide an alleviative against wage hikes, save the food industry which [was] losing its competitiveness', and mitigate the discontent of the United States and other countries seeking to export agricultural products to Japan (Tsutomu Ouchi, 1989, pp. 16–17).

One month after the July 1986 Lower-House election, in which the LDP attained 304 Lower-House seats and recovered the Lower-House majority it had lost in the December 1983 election,[5] the difficulty of altering the government's longstanding rice support program became apparent. A proposal backed by the Ministry of Finance to reduce the government-controlled producer price of rice for the first time in thirty years was rejected by the LDP during the annual discussion of by how much to adjust the producer price of rice. Prime Minister Nakasone had to intervene and side with fellow LDP members, and the government-controlled producer price of rice remained unchanged.[6]

During September 1986, both the Ministry of Agriculture, Forestry and Fisheries and the Central Union of Agricultural Co-operatives (*Zenchū*), Japan's major agricultural interest group, expressed opposition to the USTR accepting the Rice Millers' Association September 1986 Section 301 petition that called on the United States government to pressure the Japanese government to increase access to Japan's rice market.[7] While Japan's agricultural interest groups opposed increasing access to Japan's rice market, a November 1986 *Mainichi Shimbun* (newspaper) poll revealed that 48 per cent of those polled approved of rice imports, while 46 per cent opposed them.[8]

After USTR Yeutter rejected the Rice Millers' Association September 1986 Section 301 petition, Prime Minister Nakasone acknowledged that it was no longer possible to ignore domestic and international pressure, and called for the government to adopt a flexible response to the calls for change in Japan's agricultural policy. Answering questions in November 1986 from opposition party members during a Lower House Budget Committee session, Prime Minister Nakasone stated that both the producer price and consumer price of rice 'should be brought closer to international prices', and emphasized that domestic issues related to rice should be addressed through voluntary reforms that would produce social harmony.[9]

The same month that the GATT member countries agreed to the September 1986 Punta del Este Declaration, the Agricultural Policy Council, an advisory committee to Prime Minister Nakasone, issued a report entitled 'Basic Direction of Agricultural Policy Toward the 21st Century' which stated that, in the Uruguay Round negotiations, Japan 'should actively contribute to the work of establishing new rules and discipline on agricultural trade and should make its utmost effort to formulate a new worldwide order in agricultural trade'. The report stated that efforts 'should be made on import access to establish reliable and operationally effective rules that particularly provide fairer and clearer conditions for quantitative restrictions as exceptional measures' (Japan, Ministry of Agriculture, Forestry and Fisheries, 1987, pp. 13–14).

Shortly after the Uruguay Round Negotiating Group on Agriculture began meeting in Geneva during early 1987, the 'Second "Maekawa" Report', an April 1987 report of the Economic Council's Special Committee on Economic Restructuring, stated that Japan would 'take an active role in promoting the GATT Uruguay Round'. The report stated that 'Efforts to promote agricultural policy must give full consideration not only to producers, but also to consumers and the food industry.' The report called for 'Improving the existing food supply and demand system under government control by introducing competitive principles in all stages from collection to marketing.' The report stated that 'An effort should be made to reduce the differential between Japanese and overseas prices and to achieve stable foodstuff supplies at popularly acceptable prices by improving productivity and promoting imports as appropriate.'[10]

While all the lists of representatives attending the meetings of the Uruguay Round Negotiating Group on Agriculture are not available to the public, Japan's representatives in the Negotiating Group on Agriculture in 1987 included some of the country's representatives to the Office of the United Nations at Geneva, as well as representatives from the Ministry of Agriculture, Forestry and Fisheries, and from the Ministry of Foreign Affairs.[11]

During early July 1987, after the Japanese authorities agreed to the first reduction in the producer price of rice in thirty years (Schwartz, 1998, pp. 213–62), Prime Minister Nakasone, in his last policy speech before the Diet, stated 'When pondering the agricultural problem . . . it is essential to consider the importance of stable food supplies; the role that agriculture plays in environmental, employment, and spiritual and cultural aspects'. Nakasone stated that 'balanced, multifaceted agricultural policies should be flexibly and appropriately implemented in each country in keeping with individual circumstances'. The Prime Minister noted that he had succeeded in having this point incorporated into the June 1987 Venice G-7 Economic Declaration,[12] which called for giving consideration to 'food security'.[13]

During the 6–7 July 1987 Uruguay Round Negotiating Group on Agriculture meeting, responding to the United States' first major proposal to the Negotiating Group on Agriculture, Japan's representatives to the GATT stated that, although they supported further liberalization of agricultural trade, import restrictions should be allowed under specific conditions, and the special characteristics of agriculture in comparison to other traded goods should be considered.[14]

A September 1987 survey by the Prime Minister's Office was cited as revealing that 39 per cent of the respondents felt that '"at least basic foods such as rice should be produced domestically while making an effort to reduce production costs as much as possible, even though the prices [of such basic foods] are higher than foreign products"'. Another 32 per cent replied that it was '"better for Japan to produce foodstuffs while cutting production costs even though prices are higher than those of imported products"'. But, 20 per cent stated that it was '"more desirable to import foreign agricultural products if their prices [were] lower than domestic products"'. According to the Ministry of Agriculture, Forestry and Fisheries White Paper, 'Based on these results, it [could] be concluded that the majority of the Japanese desire in principle, food should be supplied domestically although it is necessary to reduce production costs.'[15]

During October 1987, a GATT dispute panel, which the United States had requested in September 1986, determined that Japan's import restrictions on twelve agricultural products were inconsistent with Japan's obligations under Article XI of the GATT, which calls for the 'General Elimination of Quantitative Restrictions' on imports. The panel ruled that Article XI was applicable, independent of whether the import restrictions 'were made effective through quotas or through import monopoly operations'.[16] Japan's Central Union of Agricultural Cooperatives and various LDP Diet members from farming areas opposed the government removing the contested import restrictions in order to comply with the GATT panel decision, and urged the government to seek some form of realistic solution through further negotiations with the United States.[17] The Japanese government believed that, of the twelve products affected by the dispute panel's decision, the liberalization of imports of dairy products and starch would have the most impact on Japan's farmers.[18]

The Takeshita Administration and the Uruguay Round Mid-Term Review Agreement

At the beginning of the Administration of Noboru Takeshita (November 1987–June 1989), during late December 1987, Japan submitted its first major proposal to the Uruguay Round Negotiating Group on Agriculture.[19] As for market access, Japan called for amendments to the GATT provisions concerning quantitative import restrictions, including the removal of waivers, such as the waiver granted to the United States in 1955 for its programs under Section 22 of the Agricultural Adjustment Act of 1933. Japan also stated that consideration should be given to the position of countries such as Japan that had low self-sufficiency rates.[20]

As a result of this October 1987 GATT dispute panel decision, the Japanese government feared increasingly that its virtual ban on rice imports might become the subject of a GATT dispute panel. Such a panel, if requested by another GATT member, would probably rule that Japan's virtual ban on rice imports was in violation of Article XI of the GATT, which called for the 'General Elimination of Quantitative Restrictions' on imports (Matsushita, 1993b, p. 286). A ruling of this nature would force Japan rapidly to remove its virtual ban on rice imports. Thus the Japanese government insisted that the liberalization of the rice market be discussed only at the multilateral level in the ongoing Uruguay Round negotiations, and refused to discuss rice import liberalization at the bilateral level with the United States.

During early February 1988, Japan informed the GATT Council that it would abide by the GATT panel report's findings concerning Japan's restrictions on twelve agricultural products. But, Japan claimed that 'the implementation of the recommendations would entail serious domestic repercussions' (GATT, *GATT Activities 1987* (1988), pp. 67–8), and refused to abolish quantitative import restrictions on two of the twelve products – dairy products and starch.[21]

During early May 1988, after the United States and Japan were unable to reach agreement on the extension of their 1984 agreement to expand Japanese beef and citrus import quotas, the United States requested the GATT Council to establish a dispute panel to determine whether Japan's import restrictions on beef, fresh oranges and orange juice were inconsistent with Japan's obligations under Article XI of the GATT. But before such a panel could be selected, Japan chose in June 1988 to conclude a new beef and citrus agreement with the United States (Odell, 2000, pp. 148–55; Porges, 1994, pp. 237–56). Under the new agreement, Japan agreed to expand its import quotas on beef significantly by April 1991, reduce its import restrictions on fresh oranges substantially by April 1991, and reduce its import restrictions on orange juice by April 1992.[22] The Japanese government entered into this agreement even though several major opposition parties, including the Japan Socialist Party, the Clean Government Party (*Kōmeitō*, henceforth Komeito), the Democratic Socialist Party, the Japan Communist Party and the Social Democratic Party jointly supported an April 1988 Diet resolution that opposed the liberalization of import restrictions on beef and oranges.[23] Several months later, the Diet approved a supplemental budget that included compensation measures for farmers who might be affected by the liberalization resulting from the agreement.

During the June 1988 G-7 Toronto Summit meeting, the Takeshita Administration emphasized that Japan would not discuss rice liberalization at the bilateral level, and that the special characteristics of Japanese agriculture should be recognized. The Takeshita Administration asserted that Japan was the only country in the G-7 that imported more agricultural products than it exported, that the country had concerns about self-sufficiency, and that the Diet had passed a resolution opposing the liberalization of Japan's rice market.[24]

According to a September 1988 *Mainichi Shimbun* poll, partly as a result of recent changes in Japanese agricultural policy, between April and early September 1988, support for the LDP fell from 70 per cent to 59 per cent among 'those involved in agriculture'. During the same period, and among the same group, support for some of the opposition parties increased, including an increase from 6 per cent to 11 per cent for the Japan Socialist Party.[25]

After the Rice Millers' Association filed its second Section 301 petition with the USTR in September 1988, a plenary session of the Upper House of the Japanese Diet passed a further resolution opposing the liberalization of Japan's rice market, as it had during the summer of 1984.[26]

During the December 1988 Montreal GATT Ministerial Uruguay Round Mid-Term Review Meeting, concern emerged within the LDP that if Japan agreed to the Uruguay Round Mid-Term Review Agreement under discussion, it might be required to make a commitment to increase import access to its rice market.[27] However, several months later, when Japan and the other GATT member countries agreed to the Uruguay Round Mid-Term Review Agreement at the April 1989 Uruguay Round Trade Negotiations Committee meeting, neither the Japanese government nor the Central Union of Agricultural Co-operatives believed that Japan had committed itself to any specific concessions concerning increased access to the Japanese rice market. Officials of the Ministry of Agriculture, Forestry and Fisheries stated that Japan was not required to make any concessions in the short term concerning access to its rice market, because the Mid-Term Review Agreement did not include a specific base year on which to calculate the conversion of import access measures into tariffs.[28] The forty-person Japanese delegation to the April 1989 Uruguay Round Trade Negotiations Committee meeting included some of Japan's representatives to the Permanent Mission to the United Nations, and representatives from the Ministry of Foreign Affairs, the Ministry of International Trade and Industry, the Ministry of Finance and the Economic Planning Agency, as well as six representatives from the Ministry of Agriculture, Forestry and Fisheries.[29]

The LDP's Agriculture, Forestry and Fisheries Trade Measures Committee, the Research Commission on Comprehensive Agricultural Policy, and the Subcommittee on Agriculture and Forestry, was reported to have issued a joint statement that the Uruguay Round Mid-Term Agreement had 'enabled the creation of a general framework for the incorporation of long- and short-term goals; [and] that negotiations [could] now be targeted at a wide range of measures, such as waivers (exemptions from the obligation to liberalize products)'. The statement also acknowledged 'that consideration of food security ha[d] been clarified, which [gave] Japan a foothold in its claims about rice; and that Japan, which [had] made efforts in reform, [could] deal with the short-term measures'. Mitsugu Horiuchi, president of the Central Union of Agricultural Co-operatives, was quoted as stating that he was '"confident that this agreement [would] have no impact"' on Japan's existing policy that Japan be completely self-sufficient in rice.[30]

The Uno Administration and the loss of the LDP Upper House majority

Shortly after becoming Prime Minister, Prime Minister Sōsuke Uno, who was Prime Minister from June to August 1989, told the Diet in a

June 1989 policy speech that he would work 'for greater productivity and ensuring stable food supplies at reasonable prices while stabilizing farm management'. But, Uno noted that Japanese agriculture faced 'a turning point in that it [was] expected to become a robustly productive industry able to withstand the harsh conditions in Japan and internationally'.[31]

In the late-July 1989 Upper-House election, the LDP lost its majority in the Upper House of the Diet, winning only 36 seats and acquiring only 109 of the 127 seats necessary for a majority in the Upper House. The Japan Socialist Party more than tripled the number of seats it held in the Upper House, from 20 to 66.[32]

One of several reasons for the fall in support for the LDP was that individual farmers began to rebel against recent reductions in government support for agriculture, including the 1987 and 1988 reductions in the producer price of rice. The government had also recently reduced its involvement in rice distribution and increased efforts to promote large-scale farming. The farmers' confidence in the LDP had been shaken further as a result of concessions the Japanese government had made in order to comply with the October 1987 GATT dispute panel decision concerning Japan's import restrictions on twelve agricultural products. Also, the Japanese government had to implement the concessions the Japanese government made in the June 1988 US–Japan beef and citrus agreement. Japanese farmers feared that more concessions, perhaps concerning the Japanese rice market, were likely to result from the ongoing Uruguay Round agriculture negotiations.[33]

The opposition parties, particularly the Japan Socialist Party, capitalized on the farmers' dissatisfaction with the LDP by opposing agricultural import liberalization, and by pledging continued government support for agriculture. In the wake of the LDP's loss of its majority in the Upper House, concerns were raised among business groups about the extent to which continuing agricultural reform might further diminish farmers' support for the LDP, and lead to greater weakening of the LDP's dominant role in the Diet. Such concerns led to a decrease in support for the agricultural policy reform agenda outlined during the Nakasone Administration.

The Kaifu Administration and the December 1990 Ministerial Meeting

During the Administration of Toshiki Kaifu (August 1989–November 1991), Japan became aware increasingly of the concessions Japan might have to make in order to conclude the Uruguay Round negotiations.

During the 25–26 September 1989 meeting of the Uruguay Round Negotiating Group on Agriculture, Japan emphasized 'the need [in a Uruguay Round agriculture agreement] to take full account of non-trade concerns, in particular food security and stable supplies of basic foodstuffs'. Japan considered that maintaining some level of domestic production was essential, since food security could not be 'ensured solely by the maintenance of potential production capacity, food stocks, bilateral agreements or diversification of suppliers'.[34] Its government representatives stated that Japan, which was highly dependent on other countries for agricultural products, had chosen to continue to provide all its own rice, and restrict imports in order to protect its own food security.[35]

During the February 1990 Lower-House election campaign, Japan's political parties competed among themselves for farmers' votes by opposing the possible liberalization of the Japanese rice market that might result from the Uruguay Round negotiations. According to the *Mainichi Shimbun*, 82 per cent of the candidates polled believed that 'not even a single grain of rice should be allowed' to be imported into Japan. Eight per cent stated that 'rice imports should be liberalized', and 'the remaining 10 per cent said that rice import liberalization should be carried out step by step'.[36]

Partially as a result of its opposition to the liberalization of the Japanese rice market during the election campaign, the LDP maintained a stable (but reduced) majority of 275 seats in the Lower House.[37] The Japan Socialist Party, which had released a national plan for agriculture during the fall of 1989 in an effort to increase its rural support, gained 53 seats to attain 136 seats in total, many of which had formerly been held by the LDP in northern Tohoku, where rice is a major industry and where there was concern about the impact of rice liberalization.[38] The other opposition parties, Komeito, the Japan Communist Party and the Democratic Socialist Party, lost seats.

Although candidates from various parties made anti-rice-market-liberalization campaign pledges during the February 1990 Lower-House election campaign, the Japanese government continued to implement one of the largest reforms in the history of Japan's staple food control system by liberalizing the domestic distribution of rice. During 1990, an auction system was introduced which handled a certain percentage of semicontrolled rice, and which represented an addition to the 1988 establishment of the non-government exchange of semicontrolled rice.

These reforms weakened the previous virtual monopoly of the National Federation of Agricultural Co-operative Associations (*Zen-Noh*) over the distribution of semicontrolled rice.[39]

During April 1990, the US International Trade Commission released a report, which the USTR had requested in September 1989, providing estimates of the tariffs that might be necessary were Japan to replace its virtual ban on rice imports with a tariff on imported rice according to the proposals for the comprehensive tariffication of agricultural products under discussion in the Uruguay Round Negotiating Group on Agriculture. The US International Trade Commission estimated that were Japan required to replace its virtual ban on rice imports with a tariff, an initial tariff rate of between 620 per cent and 733 per cent would be necessary to raise the price of rice imported into Japan to the price that domestically produced rice was currently selling for there (US International Trade Commission, 1990a, p. 4).

During June 1990, Komeito, an opposition party whose power base was in urban areas, and which had lost seats in the February 1990 election, announced that it was beginning to study possible measures that might be necessary in the event that a partial liberalization of Japan's rice market would be required in order to conclude the Uruguay Round negotiations. Komeito's announcement ended the opposition parties' previous longstanding unanimous support for Japan's virtual ban on rice imports.[40]

During early July 1990, the Chairman of the Uruguay Round Negotiating Group on Agriculture, de Zeeuw, incorporated some elements of Japan's proposals, as well as the United States' proposals, into his Framework Agreement for agriculture for the Uruguay Round negotiations. De Zeeuw's Framework Agreement called for GATT member countries to submit lists of concessions built on the parameter of the 'conversion of all border measures other than normal custom duties into tariff equivalents, irrespective of the level of existing tariffs'.[41] The inclusion of the phrase 'Relevant non-trade concerns shall be accommodated to the maximum possible extent' was considered to be an acknowledgement of Japan's concerns about food security.[42] But, the phrase was also considered to have been included in order to encourage Japan to make further concessions in the negotiations.[43] Shortly after the Framework Agreement was released, Japan objected to the inclusion of the comprehensive tariffication of agricultural products in the Framework Agreement, arguing that comprehensive tariffication would require Japan's rice market to be liberalized, which would threaten the country's food security.[44]

During the July 1990 G-7 Houston Summit, Prime Minister Kaifu attempted to increase the other G-7 leaders' understanding of Japan's lack of self-sufficiency in food. As a result, the G-7 Houston Summit Economic Declaration stated that the Uruguay Round agriculture negotiations should take into account food security.[45]

During preparations for the July 1990 Uruguay Round Trade Negotiations Committee meeting, where de Zeeuw's Framework Agreement was discussed, Japan objected to the comprehensive tariffication of agricultural products included in de Zeeuw's Framework Agreement being applied to Japan's rice market.[46] At the same time, the Central Union of Agricultural Co-operatives complained that de Zeeuw's Framework Agreement was not in accordance with Japanese Diet resolutions that called on Japan to maintain its self-sufficiency in rice.[47] But during the late-July 1990 Trade Negotiations Committee meeting, GATT Director-General Arthur Dunkel, reiterating the Houston G-7 Summit Economic Declaration,[48] described de Zeeuw's Framework Agreement as 'a means to intensify the negotiations'.[49]

In August 1990, Komeito, the second largest opposition party at the time, which earlier in the year had begun to consider some form of rice liberalization so as to win the support of urban voters,[50] further outlined its plan for the partial liberalization of rice imports in order to conclude the Uruguay Round agriculture negotiations. Komeito's proposal called for an initial annual import quota of about 200,000 tons in 1992 that would be increased by 10 per cent each year over ten years until the import quota reached 500,000 tons.[51]

During late September 1990, before the 1 October 1990 deadline set by GATT Director-General Dunkel, the Ministry of Agriculture, Forestry and Fisheries submitted to the Uruguay Round Negotiating Group on Agriculture a report on Japan's domestic protection of its agricultural markets, along with a proposed list of agricultural concessions that Japan was willing to make. While the Japanese government offered 'a 5.4 percent cut in spending on protection of three agricultural products, rice, wheat, and barley, over seven years from 1990', the Japanese government continued to assert that Japan would not comply with the conversion of non-tariff import restrictions into tariffs for staple foods, including rice and milk.[52]

During late October 1990, Japan's Ministry of Agriculture, Forestry and Fisheries released a study that indicated that, if rice liberalization occurred and all barriers to rice imports were removed, Japan's GDP would fall by 1.5 per cent, and rice output would decrease by 30 per cent.[53]

At the December 1990 Brussels GATT Ministerial Meeting, which was intended to conclude the Uruguay Round negotiations, a proposal was discussed which had been prepared by Swedish Minister of Agriculture, Mats Hellstrom, who was in charge of the negotiations concerning agriculture at the meeting. Hellstrom's proposal stated that in the 'absence of significant levels of imports' of a particular commodity, a minimum level of access from 1991–2 would be established which represented 'at least 5% of current domestic consumption of the product concerned'.[54] The proposal would have required Japan to open its rice market to imports to at least 5 per cent of its domestic consumption. As a result, Japan joined South Korea and the EC in rejecting the last-minute compromise proposal.[55]

During Spring 1991, it became clear that the liberalization of the beef and citrus markets was having much less drastic effects on Japanese agriculture than some had originally predicted.[56] Several top LDP officials, including former Ministry of International Trade and Industry Minister Kabun Mutō,[57] Takeshita faction leader Shin Kanemaru,[58] LDP Policy Affairs Research Council Chairman Takeo Nishioka,[59] and former Prime Minister Takeshita[60] came publicly to endorse or speak about some form of opening of the Japanese rice market in order to conclude the Uruguay Round negotiations. Various policy options, ranging from continuing to oppose any form of liberalization, to acceptance of some form of partial liberalization of the Japanese rice market in order to conclude the Uruguay Round negotiations, were openly discussed.

The Miyazawa Administration and the Draft Final Act

During October 1991, Prime Minister Kaifu decided he would not seek re-election, and Kiichi Miyazawa was elected prime minister. A major concern throughout the Miyazawa Administration (November 1991—August 1993) was how to work out the particular compromises that might be necessary in the event that Japan had to agree to the comprehensive tariffication of agricultural products and a minimum access commitment to agricultural markets in order to conclude the Uruguay Round negotiations.

During early November 1991, Japan's Minister of Agriculture, Forestry and Fisheries, Masami Tanabu, informed the *Nihon Keizai Shimbun* (newspaper) that ' "If we accept tariffication, we would have to amend the Staple Food Control Law. Such a bill might pass the House of Representatives (the Lower House), but it would not pass the

Opposition-controlled House of Councilors (the Upper House), making it impossible to enact."' Tanabu also noted that '"Liberalization cannot be accepted from the perspective of farmers' sentiments when rice farmers are already cutting their production."'[61]

Despite protests from Japan,[62] and a number of other countries, including South Korea, Finland, Switzerland and Mexico,[63] GATT Director-General Dunkel released a December 1991 Draft Final Act of the Uruguay Round agreements which included provisions that called for the comprehensive tariffication of agricultural products and minimum access to agricultural markets. Concerning minimum access, the 'Text on Agriculture' of the Draft Final Act stated that 'Where there are no significant imports, minimum access opportunities shall be established.' These minimum access opportunities 'shall represent in the first year of the [six- to seven-year] implementation period not less than 3 per cent of corresponding domestic consumption . . . and shall be expanded to reach 5 per cent of that base figure by the end of the implementation period'.[64]

During late December 1991, the implications of Japan agreeing to such provisions were discussed. Leaders of the LDP, Japan Socialist Party, Japan Communist Party and Democratic Socialist Party, as well as the Central Union of Agricultural Co-operatives, objected to the inclusion of the comprehensive tariffication of agricultural products in the Draft Final Act, because of its potential impact on Japan's rice market.[65] The fifteen members present at a late December 1991 Cabinet meeting reaffirmed the country's commitment to making the Uruguay Round negotiations successful, but Tanabu, Minister of Agriculture, Forestry and Fisheries, stated that he could not accept the provisions concerning the comprehensive tariffication of agricultural products that were included in the Draft Final Act.[66]

During a 1992 New Year press conference, Prime Minister Miyazawa acknowledged that Japan could not let the Uruguay Round negotiations fail.[67] During early January 1992 it was reported that aides to Prime Minister Miyazawa were studying what form of relief measures might be necessary in the event that Japan had to agree to some form of liberalization of Japan's rice market in order to conclude the Uruguay Round negotiations.[68] While Foreign Minister Michio Watanabe admitted that some form of tariffication of agricultural products might occur, one of Japan's major business federations, the Association of Corporate Executives (*Dōyūkai*), suggested that the government should quickly agree to the tariffication of rice imported into Japan.[69] But Japanese officials stated that the prospect of reaching a final Uruguay

Round agreement on agriculture in the near future was not certain because of remaining differences between the United States and the European Community.[70]

During March 1992, in accordance with the timetable and format set out in the Draft Final Act, the Japanese government presented a product-by-product list of concessions for agricultural items that it was willing to implement in order to conclude the Uruguay Round negotiations. Japan proposed to reduce domestic subsidies on rice, wheat, barley, soybeans, milk, beef, pork, chicken, eggs, vegetables and fruit by 20 per cent over six years, and to lower tariffs by an average of 30 per cent over the same period. But, Japan did not include the liberalization of import restrictions on rice and dairy products in its list of concessions, and the Ministry of Agriculture, Forestry and Fisheries stated that it would try to prevent tariffs being imposed on rice in future negotiations.[71]

As the Uruguay Round negotiations continued throughout 1992, there appeared to be growing acceptance within Japan that some form of liberalization of the Japanese rice market might be necessary to enable Japan to participate in the conclusion of the Uruguay Round negotiations. In early April 1992, during the agricultural fair where, one year earlier, the US Embassy had refused to withdraw an exhibit of illegally imported rice, the Japanese government allowed US government officials to bring in and exhibit American rice.[72]

During May 1992, the Ministry of Agriculture, Forestry and Fisheries released a new policy plan, entitled 'The Basic Direction of New Policies for Food, Agriculture, and Rural Areas', which acknowledged the need to co-exist with the global community while countering Japan's declining food self-sufficiency with measures to increase productivity, improve quality and lower the cost of production. The plan also called for the further introduction of market principles into the production and distribution of rice (Japan, Ministry of Agriculture, Forestry and Fisheries, 1992a).

During the July 1992 Upper-House election campaign, agricultural import liberalization was a less important issue than it had been in the July 1989 Upper-House and the February 1990 Lower-House election campaigns. There was a very low voter turnout, and unanimous support among the LDP, the Japan Socialist Party, Komeito, the Democratic Socialist Party, the Japan Communist Party, and the *Rengō* (union federation) Party for maintaining, for the moment, Japan's virtual ban on rice imports.[73] The LDP failed to regain a majority in the Upper House, winning only 68 of the 127 seats that were contested.

Just before the election, a new political party, which became known as the Japan New Party, was formed under the leadership of ex-Kumamoto governor Morihiro Hosokawa. In the election, Hosokawa and three other members of the Japan New Party won seats. The draft declaration of the Japan New Party that was released when the party was formed called for reform of the central government and a reduction in the influence of vested interests on the rigid political system.[74]

During the rest of 1992 and early 1993, the focus of the Uruguay Round agriculture negotiations continued to be on the resolution of differences between the United States and the EC over the agricultural export subsidy provisions in the Draft Final Act. Japan continued to wait, and at times to pressure for a breakthrough in the US–EC negotiations. During November 1992, the United States and the EC finally reached the Blair House Accord.

The Hosokawa Administration and the conclusion of the negotiations and signing of the agreements

During late July 1993, newly-appointed GATT Director-General, Peter Sutherland, who had just replaced Arthur Dunkel as Chairman of the Uruguay Round Trade Negotiations Committee, suggested a calendar by which to work out an agreement concerning access to agricultural markets. Sutherland called for a 'series of bilateral negotiations on agricultural market access in Geneva beginning 30 August through the week of 13 September' which would lead to a substantive stock-taking meeting on 15 October 1993. This would be followed by a 'submission of agreed changes and revisions to the Draft Schedules of Concessions by November 15', and a final market-access result by 15 December 1993. Japan 'entirely agreed with the Chairman['s] proposed work programme', and 'stood ready to take part in the process ahead in order to bring about a successful conclusion of the Round by the year's end'.[75]

Meanwhile, the LDP's incapacity to resolve various tensions within the Diet led to a vote of no confidence in Prime Minister Miyazawa, which forced the LDP to dissolve the Diet. In the resulting July 1993 Lower-House election, the LDP failed to retain its majority there, winning only 223 seats. Prime Minister Miyazawa indicated that he would resign. The LDP was forced to forge alliances with other parties in order to form a majority coalition so that it could remain in power. This made it possible for parties formed by dissident LDP members, such as Ichirō Ozawa's Japan Renewal Party (*Shinseitō*) and the *Sakigake* ('Harbinger', henceforth Sakigake) Party, and former opposition parties to build a majority coalition which came to include elements of the Japan Socialist Party, as well as Komeito, and the Japan New Party. This majority coalition elected Hosokawa of the Japan New Party (which held only a few seats in the Diet) as Prime Minister in August 1993.

As a result of a poor harvest, during late September 1993, the Japanese government announced that it would import 200,000 tons of rice by the end of 1993 as an emergency measure in order to counter an estimated shortage of 1 million tons for 1993.[76] But Japanese government officials stated that there was no direct relationship between this decision and the ongoing discussions of the liberalization of Japan's rice market in the context of the Uruguay Round agriculture negotiations.[77]

During early October 1993, as it became increasingly likely that the Uruguay Round negotiations might be concluded by the end of 1993, the *Mainichi Shimbun* reported that, among the major parties that made up the governing coalition, Komeito and the Japan New Party were relatively flexible concerning rice liberalization, partly because both parties were supported extensively by urban consumers. The Japan Renewal Party and Sakigake Party, which had split from the LDP, were very dependent on agricultural voters and therefore opposed liberalization. The Japan Socialist Party, since its base had become increasingly rural as a result of its recent electoral success, continued to oppose liberalization.[78]

When rumors spread in mid-October 1993 that Japan had agreed at the GATT in Geneva to the comprehensive tariffication of agricultural products and some form of minimum access to Japan's rice market, the Hosokawa Administration denied them.[79]

During a 7 December 1993 Cabinet meeting, Prime Minister Hosokawa asked the leaders of the various parties in the majority coalition to consult with their respective parties about a compromise plan that had been worked out in Geneva and concerned the liberalization of Japan's rice market. Prime Minister Hosokawa hoped to make a public announcement concerning rice liberalization by 10 December, and told the Cabinet members that there was little room to renegotiate the compromise. Under this compromise, Japan's rice market would be exempt from comprehensive tariffication for six years (in other words, Japan would not have to replace its virtual ban on rice imports with tariffs for six years), as long as it permitted a slightly higher initial access to its rice market of 4 per cent, which would grow to 8 per cent during the same period.[80]

On 8 December 1993, Tomiichi Murayama, chairman of the Japan Socialist Party, which was the largest party in the governing coalition, remained opposed to accepting the compromise proposal worked out in Geneva, because it was not in accord with the majority coalition parties' agreement that they would never lift Japan's virtual ban on rice imports.[81] The Central Union of Agricultural Co-operatives also opposed the compromise proposal under discussion, and appealed to Murayama.[82]

On 8 December 1993, it was revealed that, according to the agreement worked out in Geneva, if a country delayed the implementation of comprehensive tariffication of agricultural products for more than six years, that country would be forced to make additional concessions.[83] While this made it more difficult for Japan to accept the compromise plan, all the political parties in Prime Minister Hosokawa's coalition eventually accepted the Prime Minister's decision, apart from the Japan Socialist Party, which opposed the decision, but expressed understanding.[84]

On 14 December 1993, the last day of the Uruguay Round negotiations, Prime Minister Hosokawa announced Japan's agreement to partially liberalize the Japanese rice market in order to conclude the GATT Uruguay Round negotiations.[85] Prime Minister Hosokawa emphasized that the Japanese government had achieved an agreement to postpone tariffication for six years, and stated that the agreement did not contradict earlier Diet resolutions that opposed the liberalization of Japan's rice market. The chairman of the Federation of Economic Organizations (*Keidanren*) welcomed the decision,[86] but the Central Union of Agricultural Co-operatives opposed it.[87] Some opposition LDP members also were against the Japanese government's decision to partially liberalize the Japanese rice market.[88]

On 15 December 1993, the Japanese delegation as well as the delegates of the other GATT member countries approved the Final Act of the Uruguay Round negotiations, including the Agreement on Agriculture. According to the GATT, in order 'to facilitate the implementation of tariffication in particularly sensitive situations', such as Japan's virtually closed rice market, a '"special treatment"' clause had been introduced into the Uruguay Round Agreement on Agriculture. This special treatment clause allowed a country 'under certain carefully and strictly defined conditions' to maintain import restrictions up to the end of a six-year implementation period. The clause required such countries to agree to greater minimum access commitments than those set out in the December 1991 Draft Final Act. Such countries would have to establish 'minimum access opportunities [that] represent[ed] four per cent of domestic consumption of the designated products in the first year of the implementation period [which would] increase annually to reach 8 per cent in the sixth year [of implementation]'. The Final Act stated that 'the final [import percentage] figure [would be] lower if the designated products [were] tariffied before the end of the implementation period'. 'In case of any continuation beyond the sixth year, additional commitments [would] have to be taken.'[89]

A few days later, the Ministry of Agriculture, Forestry and Fisheries announced that, in accordance with the Final Act, effective from fiscal 1995, Japan would set the import quota for rice for 1995 at 379,000 tons, or 4 per cent of its annual consumption. This quota would be expanded by 0.8 per cent over the following six years, to reach 8 per cent of annual consumption, or 758,000 tons, by the year 2000. The Ministry of Agriculture, Forestry and Fisheries stated that, during these six years, imports of rice would be duty-free, and the importation of a number of other farm products would be affected by tariffication and tariff reductions.[90]

During mid-April 1994, towards the end of the Hosokawa Administration, Japan signed the Uruguay Round agreements, including the Agreement on Agriculture, in Marrakesh, Morocco. Annex 5, which concerns Article 4(2) of the Uruguay Round Agreement on Agriculture, stated that for some products, such as Japanese rice, quantitative import restrictions would not have to be converted into ordinary customs duties during the six-year implementation period beginning in 1995 provided certain conditions were met. Imports had to comprise 'less than 3 per cent of corresponding domestic consumption in the base period 1986–1988'. Also, minimum access opportunities had to correspond to '4 per cent of base period domestic consumption' 'from the beginning of the first year of the implementation period and, thereafter, [be] increased by 0.8 per cent of corresponding domestic consumption in the base period per year for the remainder of the implementation period'.[91]

The Hata Administration, the Murayama Administration, and the passage of implementing legislation

After Prime Minister Hosokawa resigned in April 1994, the eight-month-old seven-party coalition that had supported Prime Minister Hosokawa elected Tsutomu Hata of the Japan Renewal Party as prime minister. Hata defeated an LDP candidate and a Japanese Communist Party candidate.[92] But, the coalition that supported Prime Minister

Hata soon fell apart because of tensions between the Japan Socialist Party and Hata and Ozawa's Japan Renewal Party.

During late June 1994, the LDP submitted a motion of no confidence in Prime Minister Hata, and he left office after fifty-nine days. As alliances and parties within the Diet continued to change rapidly, the Lower House of the Diet elected Japan Socialist Party Chairman, Tomiichi Murayama, as Prime Minister in a run-off vote. Murayama led a coalition that included his own party, the LDP and the Sakigake Party.[93]

On 2 December 1994, the Lower House approved the Uruguay Round agreements, including the Agreement on Agriculture and seven related bills including the Law for Stabilization of Supply-Demand and Price of Staple Food (which replaced the Food Control Law). The Upper House passed the Uruguay Round agreements and related legislation into law on 8 December 1994. Among other measures included in the Uruguay Round implementing legislation, was a six-year, 6.01 trillion yen ($60.7 billion) package of domestic farm measures.[94]

As one scholar notes, when the Japanese Cabinet submitted the Uruguay Round agreements and implementing legislation to the Diet for approval in late 1994, 'most of the opposition parties were not in a position to criticize the Government for proposing ratification' of the agreements. This was because most of the parties in opposition in December 1994 were part of the coalition that had supported Prime Minister Hosokawa in December 1993, and were represented in the Cabinet when Japan accepted the Final Act and the Agreement on Agriculture at the close of the Uruguay Round negotiations. Thus, only the Japan Communist Party – the one party that was neither a member of the coalition in December 1993 that supported Prime Minister Hosokawa, nor a member of the coalition in December 1994 that supported Prime Minister Murayama – was able to oppose effectively the ratification of the Uruguay Round agreements.[95]

Domestic politics and international relations in Japanese trade policymaking during the negotiations

Based on the preceding discussion of Japanese trade policymaking during the Uruguay Round agriculture negotiations, the validity of the three hypotheses discussed in Chapter 2 concerning the extent to which domestic politics affects the agreement that Japan reaches in a particular trade negotiation can be assessed.

As the first hypothesis suggests, because key elements within the Japanese bureaucracy and a large number of Diet members favored the Japanese government adopting a protectionist stance during the Uruguay Round agriculture negotiations, Japanese negotiators were able to oppose a Uruguay Round agriculture agreement that included the comprehensive tariffication of agricultural products and a commitment to minimum access to agricultural markets until the very end of the Uruguay Round negotiations.

While Prime Minister Nakasone advocated the initiation of the Uruguay Round negotiations, and Japan's Agricultural Policy Council had hoped through the negotiations 'to establish reliable and operationally effective rules that particularly provide fairer and clearer conditions for quantitative restrictions as exceptional measures',[96] the July 1990 Framework Agreement of the Chairman of the Uruguay Round Negotiating Group on Agriculture, de Zeeuw, included provisions concerning the comprehensive tariffication of agricultural products which might require Japan to commit to more of a liberalization of the Japanese rice market than some elements of the bureaucracy and members of the Diet intended at the beginning of the negotiations.

Within the bureaucracy, tensions existed inside and between the Ministry of Agriculture, Forestry and Fisheries, the Ministry of International Trade and Industry, and the Ministry of Foreign Affairs concerning the negotiating stance that Japan should adopt in the Uruguay Round agriculture negotiations.

One of the driving forces behind domestic agricultural reform within the Japanese bureaucracy was the Ministry of Finance, which had been pressuring the Ministry of Agriculture, Forestry and Fisheries since the late 1960s to reduce deficitary spending on Japan's costly food control system (Hideo Sato and Schmitt, 1993, p. 272).

During the negotiations, the Ministry of Agriculture, Forestry and Fisheries wavered at times in implementing the major tenets of the agricultural policy set out in the Ministry's 1986 report, 'Basic Direction of Agricultural Policy Toward the 21st Century'. While the Ministry of Agriculture, Forestry and Fisheries was chiefly responsible for designing and implementing agricultural policy, the Ministry was not very skillful at gathering public support for the reforms it sought to undertake, or in building a domestic consensus behind Japan's negotiating stance at the Uruguay Round agriculture negotiations. There was a reluctance by the Ministry to produce and encourage the public debate of accurate, realistic calculations concerning the impact on Japan of the partial liberalization of its rice market, and to set up some sort of coordinating mechanism to reduce the polemic debate that was occurring among bureaucrats, political leaders and farmers concerning Japan's negotiating stance in the Uruguay Round agriculture negotiations.

The Ministry of International Trade and Industry came to favor some form of partial liberalization of Japan's rice market in order to speed the conclusion of the Uruguay Round negotiations, in the hope that the liberalization brought about by such negotiations would promote domestic economic growth, while the Foreign Ministry sought to promote the image of Japan cooperating actively in international affairs, and became involved in the debate over rice liberalization only to the extent that it was relevant to Japan's foreign relations.

In the end, the Japanese bureaucracy skillfully reached the conclusion of the Uruguay Round negotiations without giving up too much in terms of agricultural concessions. Reform-minded bureaucrats of the Ministry of Agriculture, Forestry and Fisheries and the Ministry of Finance made some progress in conducting a difficult agricultural reform by agreeing to a partial liberalization of Japan's rice market, but the Ministry of International Trade and Industry lost out somewhat because the delayed conclusion of the Uruguay Round probably impeded Japanese economic growth during the early 1990s, although the timing of the conclusion of the Uruguay Round negotiations was far from being entirely in Japan's control.

During the negotiations, the Japanese bureaucracy came to be more divided and less allied to the LDP than it had been during the previous two decades. Many Diet members, as well as bureaucrats within the Ministry of Agriculture, Forestry and Fisheries, were concerned with protecting what remained of Japan's agricultural self-sufficiency and opposed to making any concessions concerning Japan's virtual ban on rice imports until the end of the Uruguay Round negotiations.

As the negotiations wore on, the more internationally-oriented ministries, including the Ministry of International Trade and Industry, and the Ministry of Foreign Affairs, as well as certain factions of the LDP, joined by members of various political parties, including Komeito and the Japan New Party, came to advocate some form of partial liberalization of Japan's rice market in order to promote the conclusion of the Uruguay Round negotiations.

As the second hypothesis suggests, the more divided the Japanese government became concerning the negotiating stance Japan should take in the Uruguay Round agriculture negotiations, the more influence the Japanese Diet had over the terms of the negotiated agreement, and the more likely the Japanese government was to alter its negotiating stance in order to reach an agreement.

Since the Diet had to ratify the results of the Uruguay Round negotiations and pass legislation to implement the results of the negotiations, the bureaucracy had to craft policies during the Uruguay Round agriculture negotiations that would secure the endorsement of the Diet. The compromise worked out, the special treatment clause in Annex 5 concerning Article 4(2) of the Uruguay Round Agreement on Agriculture that allowed Japan to delay tariffication for six years if Japan agreed to a slightly higher minimum access rate to its rice market, resulted from Japanese negotiators having insisted continually that the Diet might not ratify a Uruguay Round agreement that would require Japan to liberalize its rice market.

Some Diet members within the anti-rice-market-liberalization coalition were dependent on farming constituencies that might be affected by such liberalization. These Diet members allied themselves with bureaucrats within the Ministry of Agriculture, Forestry and Fisheries who were being affected negatively by the ongoing reforms of Japan's food control system.

Diet members within the pro-rice-market-liberalization coalition dependent on urban voters allied with the internationally-oriented bureaucracies – the Ministry of International Trade and Industry and the Foreign Ministry – and advocated the liberalization of the Japanese rice market in order to conclude the Uruguay Round negotiations.

To the extent that the Diet could influence the bureaucracy concerning the negotiations, influence shifted to the Diet, and Japan's negotiating stances at the Uruguay Round agriculture negotiations became closer to the Diet's preferences. These preferences were reflected in the 1984 and 1988 Diet resolutions expressing opposition to the liberalization of Japan's rice market, and in various statements during the negotiations by Diet members concerning the negotiating stance Japan should adopt.

The fear that the Japanese Diet would not accept the Uruguay Round agriculture agreement reached in Geneva increased the Diet's influence over the terms of the negotiated agreement, and was one of the key reasons for Japanese negotiators' insistence until the end of the negotiations that Japan should receive exceptional treatment for its rice market.

The general decline and diminishing legitimacy of the LDP, changing leadership in the Diet, and a series of relatively short prime ministerships marked the 1986–94 period. There were greater, more fundamental, divisions within the LDP, as the party strove to carry out various reforms and to respond to the changing nature of the Japanese electorate (Pempel, 1993, pp. 124–9). While the LDP lost seats in the election of December 1983, it gained a fresh mandate in the double election of July 1986. However, a series of scandals, from the Recruit to the Sagawa Kyūbin scandal, weakened the LDP's leadership. As Curtis (1999, pp. 86–8) describes, in August 1992, after allegations of having received unreported funds from the Sagawa Kyūbin company, a package delivery company, Shin Kanemaru resigned as vice president of the LDP and then later resigned as head of the Takeshita faction. In the wake of the scandal, as Hisane Masaki noted, Prime Minister Miyazawa could 'no longer count on the enormous political influence' wielded by Kanemaru and Takeshita, who both had come to advocate a more flexible stance concerning rice liberalization, 'to build party consensus on such a politically sensitive issue as rice'.[97] The LDP's power was also weakened as a split emerged between LDP Diet members who staunchly defended Japan's virtual ban on rice imports, and LDP Diet members more dependent on urban voters who advocated some form of liberalization of Japan's rice market in order to conclude the Uruguay Round negotiations. An increasing number of rural votes that might have been cast for the LDP were cast for the Japan Socialist Party which opposed any form of liberalization of Japan's rice market. Also, an increasing number of urban voters who might have voted for the LDP instead voted for parties such as Komeito that were openly discussing rice liberalization. The end result was the splintering of the LDP, culminating in the emergence of the new spin-off parties, such as the Japan Renewal Party and the Sakigake Party, and the selection of the first non-LDP prime ministers, Hosokawa and Murayama, for the first time in forty years.

While divisions between the bureaucracy and the Diet may have increased the Diet's influence on the negotiations, the more divided the Diet became concerning rice liberalization, the less easy it was for the Diet to take unified stances opposing the liberalization of Japan's rice market, and the less weight the Diet's anti-rice-market-liberalization resolutions carried. Gradually, however, opposition within the bureaucracy and the Diet to making concessions concerning the liberalization of Japan's rice market eroded during the course of the negotiations, making it easier at the end of the negotiations for the Japanese government to agree partially to liberalize Japan's rice market.

As for the validity of the third hypothesis, because there were a substantial number of informed domestic groups in Japan who supported

Japan agreeing to partial liberalization of its rice market in order to conclude the Uruguay Round negotiations, it was possible for the Japanese Diet to ratify a Uruguay Round agreement requiring Japan to liberalize. Because such groups existed, it was easier for other countries to believe that Japan might cooperate in reaching a Uruguay Round agriculture agreement.

Japan's negotiating stance during the Uruguay Round agriculture negotiations were developed largely within the bureaucracy, particularly in the Ministry of Agriculture, Forestry and Fisheries, but was formed in response to ongoing pressure from other ministries, political parties and interest groups. When the LDP lost its majority in the Upper House in the July 1989 Upper-House elections, the consensus in support of agricultural reform that existed at the beginning of the Uruguay Round negotiations weakened among various bureaucrats of the Ministry of Finance, the Ministry of Agriculture, Forestry and Fisheries, and the Ministry of International Trade and Industry, business- and consumer-oriented Diet members, and businessmen.

During the negotiations, a fissure appeared in what Pempel calls the 'alliance of iron and rice', the 'fusion of two key social sectors: business and agriculture', on which the LDP's dominant conservative coalition rested for most of the postwar period (Pempel, 1993, p. 108). This consensus between business and agriculture, and between the agricultural cooperatives and farmers concerning rice support programs and LDP rule, was shattered as domestic pressure mounted to improve agricultural productivity and as international calls for greater access to Japan's markets increased. Polemic debates concerning the possible impact of the liberalization of Japan's rice market occurred within and between business and agricultural interest groups, and among agricultural cooperatives and farmers, weakening longstanding alliances between them, which led to the decline in power of the LDP. As a result, the LDP could not manage to achieve the agricultural reforms that had been outlined during the Nakasone Administration without weakening significantly the power base of LDP members dependent on rural constituents.

Some Diet members dependent on rural constituents, bureaucrats within the Ministry of Agriculture, Forestry and Fisheries who opposed ongoing reforms of Japan's food control system, the agricultural cooperatives, most farmers, and consumer organizations had supported Japan's virtual ban on rice imports for decades.[98] The main objective of the members of this anti-rice-market-liberalization coalition was to minimize the concessions Japan would have to make concerning its virtual ban on rice imports in order to conclude the Uruguay Round negotiations.

The pro-rice-market-liberalization coalition that emerged during the negotiations included reform-minded bureaucrats within the Ministry of Agriculture, Forestry and Fisheries, and bureaucrats within the Ministry of International Trade and Industry, the Ministry of Foreign Affairs, and the Ministry of Finance, as well as LDP Diet members dependent on the votes of businessmen and consumers, Japanese business federations, some of the media, and some consumers and farmers. Internationally-oriented bureaucrats, Diet members and businessmen were particularly concerned that Japan's delay in making concessions concerning the liberalization of its rice market might increase anti-Japanese sentiment in the United States and have an adverse impact on Japan's foreign relations with other countries. Business groups generally sought the support of the Ministry of International Trade and Industry in promoting agricultural reform and the conclusion of the Uruguay Round. By the time business groups began to realize the danger to LDP dominance of promoting agricultural reform, the party's dominance had already been damaged by the results of the July 1989 Upper-House election. As economic growth slowed in the late 1980s and early 1990s, the consensus in favor of domestic agricultural reform that had existed at the beginning of the Uruguay Round weakened. Nonetheless, business groups, such as the Federation of Economic Organizations, continued to favor increasing the efficiency and productivity of Japanese agriculture, and some form of liberalization of the Japanese rice market in order to conclude the Uruguay Round negotiations, and came to do so more as the recession of the early 1990s worsened. As the pro-liberalization coalition became stronger, it became more likely that the Diet would ratify an agreement that included some form of liberalization of the rice market, and became easier for other countries to believe that Japan might cooperate in reaching a Uruguay Round agriculture agreement.

The outcome of the GATT dispute panel concerning Japan's import restrictions on twelve agricultural items and the outcome of the beef and citrus negotiations with the United States led to reductions in government support for agriculture which were unpopular among individual farmers. These reductions in government support were a key reason for the LDP's inability to retain, after July 1989, its majority in the Upper House of the Diet. The factions within the LDP that promoted agricultural reform appear to have misjudged the transition in the electorate from farmers to urban consumers, underestimating the importance of farmers to LDP dominance. The LDP appears not to have prepared Japan's farmers adequately for such reforms, or debated compensation measures while responding to international pressure for the liberalization of Japan's agricultural import restrictions.

After losing its majority in the Upper House in the July 1989 Upper-House election, the LDP responded to the farmers' discontent by joining the other political parties in openly opposing liberalization of Japan's rice market during the February 1990 Lower-House election. Partly as a result of this action, the LDP was able to maintain a stable majority coalition in the Lower House. But, two years later, for a variety of reasons, the LDP was unable to regain its majority in the Upper House in the July 1992 Upper-House election. As a result of the LDP not winning enough seats in the July 1993 Lower-House election, a majority coalition was formed by dissident LDP elements and former opposition parties which selected a non-LDP Diet member, Hosokawa, as Prime Minister (Curtis, 1999, pp. 99–114).

Agricultural reform, including the liberalization of Japan's rice market, was possible if the influence of the rural-vote-dependent Diet members on Ministry of Agriculture, Forestry and Fisheries bureaucrats could be reduced. However, weakening the farmers' ties to the LDP was difficult, and getting some of the bureaucrats in the Ministry of Agriculture, Forestry and Fisheries to cooperate in the dismantling of the longstanding food control system was not easy.

Japan's major agricultural interest group, the Central Union of Agricultural Co-operatives, actively opposed any form of liberalization of the rice market throughout the Uruguay Round negotiations, although the ongoing liberalization of the domestic rice distribution system gradually reduced the influence of the Central Union of Agricultural Co-operatives on the Ministry of Agriculture, Forestry and Fisheries. Opposition parties, such as the Japan Socialist Party and the Japan Communist Party, continued to try to use the unpopularity of agricultural reform among farmers to increase rural support for their parties and to weaken the LDP. However, the opposition parties' previous united opposition to rice liberalization was shattered after Komeito began to explore during 1990 the possibilities of the partial liberalization of the Japanese rice market.

Within the anti- and pro-liberalization coalitions, Japanese farmers and business groups played important roles, acting as pressure groups and information providers. As the pro-liberalization coalition gradually became more influential, and an increasing number of Diet members came to acknowledge the importance of the liberalization of the

Japanese rice market to the conclusion of the Uruguay Round negotiations, the likelihood increased that the Diet would ratify an agreement that included some form of liberalization of the rice market.

Although Japan's decision to agree to some form of liberalization of the rice market in order to conclude the Uruguay Round negotiations was met by some protest,[99] the concessions Japan made concerning the liberalization of rice imports had been debated long enough to enable many in the anti-liberalization coalition to see the importance of making some form of compromise in order to conclude the negotiations. Therefore, the Diet was able to endorse the results of the negotiations by ratifying the Uruguay Round agreements and passing implementing legislation during late 1994.

5
US–Japan Trade Policymaking during the Uruguay Round Agriculture Negotiations

Building on the discussion in Chapters 3 and 4, this chapter examines US–Japan trade negotiations during the Uruguay Round agriculture negotiations, focusing on those concerning market access. The chapter concludes by addressing the question: To what extent did domestic politics affect the agreement the United States and Japan reached in the Uruguay Round agriculture negotiations? The question is answered by assessing the validity of the three hypotheses concerning US–Japan trade negotiations set out at the end of Chapter 2.

The GATT Committee on Trade in Agriculture

Because of mounting fiscal difficulties resulting from the Common Agricultural Program in the EC and from agricultural subsidies in Japan, the EC and Japan were more willing to place comprehensive fundamental agricultural trade reform on the negotiating agenda of the Uruguay Round than they had been during the GATT Kennedy and Tokyo Rounds. During the GATT Kennedy Round negotiations (1963–7), the first systematic discussion of non-tariff trade barriers and their possible negotiation within the GATT occurred. The GATT Tokyo Round negotiations (1973–9) resulted in not only concessions concerning specific tariffs and import quotas, but also the establishment of behavioral codes and international agreements concerning dairy products, beef and general agricultural policy (James Houck, 1982, pp. 68–70, 75).

The November 1982 GATT Ministerial Meeting Declaration called for the examination of a number of matters related to trade in agriculture. Such matters included 'Trade measures affecting market access and supplies, with a view to achieving greater liberalization in the trade of agricultural products, with respect to tariffs and non-tariff measures, on a basis of overall reciprocity and mutual advantage under the General Agreement' ('Ministerial Declaration', GATT, *GATT Activities 1982*, 1983, pp. 17–18).

During the early 1980s, the United States and Japan were among the principal proponents of a new GATT negotiating round that would lead to realizing the work program described in the November 1982 GATT Ministerial Meeting Declaration. Prime Minister Nakasone stressed 'the importance of promoting the preparations of a new round of multilateral trade negotiations in order to consolidate the free trading system and to inject renewed confidence in the world economy' during President Reagan's November 1983 visit to Japan. President Reagan added that 'Progress in Japan–U.S. trade issues can foster greater trade liberalization efforts worldwide, such as the Prime Minister's call for a new round of multilateral trade negotiations, which I heartily endorse.'[1]

By early 1984, the sixty-member GATT Committee on Trade in Agriculture established in the November 1982 GATT Ministerial Declaration had issued a number of recommendations, including a call for the elaboration of conditions '"under which substantially all measures affecting agriculture would be brought under more operationally effective GATT rules and disciplines"'. A major concern of the GATT Committee on Trade in Agriculture was how to improve access to member countries' markets, and in particular the reduction of 'quantitative restrictions and other related measures affecting imports' (GATT, *GATT Activities 1984*, 1985, p. 11). By June 1984, the committee was examining specific existing forms of quantitative restrictions on imports, including the GATT waiver that the United States had received in 1955 for import quotas in existence as a result of Section 22 of the Agricultural Adjustment Act of 1933. The United States, the EC, Australia, New Zealand, Canada, Argentina and Chile supported the recommendations under discussion, but Japan was less enthusiastic about them.[2]

During the June 1984 London Seven-Nation Economic Summit, Japan called for a new GATT negotiating round, suggesting that preparations be made during 1985, and negotiations begin in 1986.[3] While the United States and Canada also supported starting a new negotiating round, France opposed initiating one too quickly. The communiqué issued at the end of the summit called on the leaders of the seven nations, 'building on the 1982 GATT work programme, to consult partners in the GATT with a view to decisions at an early date on the possible objectives, arrangements and timing for a new negotiating round'.[4]

During March 1985, a committee of seven independent experts selected by GATT Director-General, Arthur Dunkel, issued the 'Leutwiler Report'. While one of the experts was US Senator Bill Bradley, none of the experts were from Japan. The committee set out fifteen proposals for action to be taken during the next GATT negotiating round, the second of which was that 'Agricultural trade should be based on clearer and fairer rules, with no special treatment for particular countries or commodities. Efficient agricultural producers should be given the maximum opportunity to compete' (GATT, *Trade Policies*, 1985, p. 9). During 1985, a report to the Trilateral Commission called for 'more market-oriented' domestic agricultural programs and for the trilateral countries to 'move together in achieving more market-oriented policies for agriculture', stating that 'as a general guideline, a period of five to seven years should be adequate for most of the necessary adjustments'. The report stated that 'the minimum objective of [the new GATT] negotiations on the import side should be to bring about a higher degree of observance of [the GATT Article XI] provisions, including those with respect to minimum access' (D. Gale Johnson *et al.*, 1985, pp. 45–6, 52–3).

The Preparatory Committee

During January 1986, the Preparatory Committee for the upcoming GATT negotiating round began to meet under the chairmanship of Arthur Dunkel. Focusing on problems that had been highlighted in the November 1982 Ministerial Meeting Declaration, the Committee set out 'to determine the objectives, subject matter, modalities for and participation in' the next round of GATT negotiations. The Committee hoped to prepare, by mid-July 1986, recommendations for a negotiating agenda for the forthcoming negotiations (GATT, *GATT Activities 1985*, 1986, p. 10).

During April 1986, a US Department of Agriculture official announced the United States' objectives concerning agriculture in the forthcoming round of GATT negotiations: 'First, to improve access to foreign markets; second, to write effective GATT rules for controlling unfair trade practices such as subsidization; and, third, to harmonize the use of food, plant and animal health restrictions which impede trade.'[5] During the same month, Japan presented a proposal to the Preparatory Committee, expressing concern about the negotiations concerning agriculture that were taking place in the Committee.[6] An April 1986 annual 24-nation OECD meeting communiqué called for agriculture to be included in the agenda of the next GATT round.[7] During a June 1986 meeting of nineteen trade ministers in Seoul, Korea, several agricultural exporting countries stated that the upcoming GATT round should include discussions of the liberalization of agricultural trade.[8] A few weeks later, the twelve member countries of the EC agreed to discuss agricultural subsidies during the forthcoming round of GATT negotiations.[9]

By early August 1986, there were three main drafts of the negotiating agenda for the forthcoming GATT negotiating round. The majority-backed draft was supported by about forty countries, including the United States, the European Community and Japan. Another draft was supported by about ten developing countries.[10] One of the many points of disagreement was agriculture. The United States and Australia called for a reduction in agricultural export subsidies. The EC, responding to opposition from France, opposed such a reduction.[11] During an early September 1986 quadrilateral meeting – including the United States, the EC, Japan and Canada – the EC blocked the endorsement of the majority-backed draft.[12] This draft, which, according to one source, by then had the support of about thirty-two countries, approximated the United States' position, but met with opposition from France (and therefore also the EC) because it included references to ' "increasing discipline on the use of all subsidies affecting agricultural trade" '.[13]

The Punta del Este Declaration

Shortly before the September 1986 Punta del Este GATT Ministerial Meeting during which the negotiating agenda for the Uruguay Round of negotiations was established, Japan stated that, while Japan had been participating in the discussions concerning possible reductions in protectionist policies for agricultural products thus far, it was in favor of a slower implementation of agricultural reform than was advocated by the United States and other countries.[14] Since Japan did not export many agricultural products, it was most concerned about the potential impact of the negotiations on its remaining import restrictions on twenty-two agricultural products.[15]

According to a United States government official who attended the September 1986 Punta del Este GATT Ministerial Meeting, the United States 'found itself in the middle' in the debates that took place. 'On one side there was a group known as G-14 [the Cairns Group], the group of 14 nations which were primarily the nonsubsidizing agricultural exporters who wanted a very, very tough agenda with respect to agricultural negotiations – countries like Australia, Argentina, Canada. And on the other side was the Common Market led by the French who in essence wanted to water down the process so that they could continue to do what they wanted.'[16]

Japan, under pressure to open its markets in order to reduce its growing trade surplus with many countries, supported the majority-backed proposal that had received the support of fifty-nine countries.[17]

The Punta del Este Declaration that resulted from the September 1986 GATT Ministerial Meeting closely resembled the majority-backed draft that had emerged from the Preparatory Committee.[18] The section of the Declaration concerning access to foreign agricultural markets stated that the negotiations would 'aim to achieve greater liberalization of trade in agriculture and bring all measures affecting import access and export competition under strengthened and more operationally effective GATT rules and disciplines, taking into account the general principles governing the negotiations, by (i) improving market access through, *inter alia*, the reduction of import barriers'.[19]

During the Punta del Este meeting, Japan's Foreign Minister Tadashi Kuranari met with USTR Yeutter and US Agriculture Secretary Richard Lyng and urged the United States government to reject the Section 301 petition that the Rice Millers' Association had recently filed with the USTR. Kuranari emphasized the importance of rice to Japanese agriculture and the political sensitivity of altering Japan's virtual ban on rice imports.[20]

During October and November 1986, the Uruguay Round Trade Negotiations Committee met for the first time. According to the Punta del Este Declaration, the Trade Negotiations Committee was to supervise the Uruguay Round negotiations. The United States' representatives to the first meetings of the Trade Negotiations Committee included officials from the USTR's offices in Geneva and Washington, and from the US Department of Commerce. Japan's representatives included Japanese representatives to the Office of the United Nations at Geneva, and representatives from the Ministry of Foreign Affairs, the Ministry of International Trade and Industry, and the Ministry of Finance.[21]

The first negotiating proposals

During late January 1987, the negotiating structure for the Uruguay Round negotiations was put in place. Fourteen separate negotiating groups, including one called the Uruguay Round Negotiating Group on Agriculture, were established. The work of the Negotiating Group on Agriculture was reviewed periodically by the Group of Negotiations on Goods, which was supervised, along with the Group of Negotiations on Services, by the Trade Negotiations Committee. GATT Director-General Dunkel chaired the Trade Negotiations Committee. Each negotiating group had its own chairperson. The Negotiating Group on Agriculture was initially to identify major problems and their causes, consider 'basic principles to govern world trade in agriculture', and submit and examine participating countries' proposals that aimed toward achieving the negotiating objectives set down in the Punta del Este Declaration.[22]

Aart de Zeeuw of the Netherlands, who had chaired the GATT Committee on Trade in Agriculture, was appointed Chair of the Negotiating Group on Agriculture.[23] This group met every few months at the GATT in Geneva until late 1990.[24] In early 1991, the entire negotiating structure of the Uruguay Round was rearranged, and Dunkel assumed control of the negotiating group that dealt with agriculture.

During the meetings of the Negotiating Group on Agriculture, countries presented various proposals concerning the agreement that should be reached, and the concessions their country was willing to make. More than thirty countries presented at least one proposal to the Uruguay Round Negotiating Group on Agriculture. Groups of countries, like the Cairns Group of agricultural exporting countries and the Nordic countries, also presented proposals. Certain international organizations, such as the Food and Agriculture Organization of the United Nations, were allowed to attend the meetings of the Negotiating Group on Agriculture as observers, on the understanding 'that the rules on the confidentiality of the proceedings of the negotiating bodies will also apply to such representatives and that the content of discussions and documents would not be available for use outside international secretariats attending meetings'.[25]

Much of the consensus concerning liberalizing agricultural trade that came to be included in the Uruguay Round Agreement on Agriculture was developed not only in the meetings of the Negotiating Group on Agriculture, but also in other Uruguay Round Negotiating Groups, and in annual G-7 summits, OECD meetings and other plurilateral meetings.

The progress of the negotiations was affected by the evolution of US–EC, US–Japan, and other countries' trade relations. Leadership in creating the consensus that emerged came not only from the United

States, the EC and Japan, but also from other countries (and groups of countries) such as the Cairns Group. Leadership also came from the GATT Director-General and other GATT officials, including the Chairman of the Negotiating Group on Agriculture, particularly when a compromise needed to be struck in order to move the negotiations forward.

Documents related to the Uruguay Round negotiations have recently become publicly available on microfiche.[26] But, at the time of writing, many of the GATT documents related to the negotiations are still restricted, and many of the government documents related to the development of the negotiating stances of the United States and Japan are not publicly available. Therefore, this book is based on derestricted GATT documents and information gleaned from other publicly available contemporaneous sources. While the negotiations received much global media attention, many of the GATT Uruguay Round documents cited in this chapter have only recently been made available to the public.

The major debate within the Negotiating Group on Agriculture occurred between the United States and the EC, and concerned reaching a mutual agreement to reduce agricultural export subsidies (Breen, 1993, pp. 168–228). Another major debate within the Negotiating Group on Agriculture concerned reaching a mutual agreement about access to foreign agricultural markets. Some of the most active participants in this debate were the United States and Japan. As stated earlier, this book focuses on this market access debate, which involved many more countries than the United States and Japan, and which was one of a large number of issues addressed by the Negotiating Group on Agriculture.

During late March 1987, the GATT Secretariat, in consultation with the Chairman of the Negotiating Group on Agriculture, outlined the major problems affecting agricultural trade, mentioning 'highly restrictive import barriers' as one of the specific problems common to the major traded commodities.[27]

By the time of the April 1987 quadrilateral meeting in Kashikojima, Japan, it was apparent that while Canada generally sided with the United States and sought a Uruguay Round agreement that would promote rapid agricultural liberalization, Japan supported the EC in advocating a more gradual plan for agricultural reform.[28] It was also evident that one of the key issues that would have to be negotiated between the United States and Japan during the Uruguay Round agriculture negotiations was the possible liberalization of the Japanese rice market.[29]

During April 1987 negotiations in Japan between US Secretary of Agriculture, Richard Lyng, and Japan's Minister of Agriculture, Forestry and Fisheries, Mutsuki Katō, Secretary Lyng requested that Japan provide some access to Japan's rice market.[30] Katō objected to discussing rice at the bilateral level, emphasizing that Japan and the United States should not decide in advance, bilaterally, that rice should be one of the topics discussed at the Uruguay Round.[31] Katō stated that, if at an advanced stage of the Uruguay Round the issue of rice were to be discussed, Japan would seek GATT's understanding on the special importance of rice and Japan's national policy of maintaining a self-supply of this commodity.[32] Upon returning to the United States, Secretary Lyng is reported to have denied that it had been his intention to ' "zero-in" ' on the sensitive rice issue. Lyng noted that Japan's insistence on maintaining its virtual ban on rice imports only pointed to Japan's ' "clear symbolic position of extreme protectionism" '.[33]

A few days later, in response to questions from Japan's *Asahi Shimbun* (newspaper), President Reagan stated that 'An open market would be in the interest of the Japanese consumer and the world trading community. I want the GATT negotiations on agriculture to be comprehensive. We have said that we are willing to put everything on the table, but we expect other countries to do the same.' President Reagan continued, 'We have an opportunity to resolve problems in agricultural trade which have been an economic drain on many countries. In order to solve these problems, we must all cooperate, and we must all be willing to put our agricultural programs and policies on the negotiating table.'[34]

During the 5–6 May 1987 meeting of the Uruguay Round Negotiating Group on Agriculture, basic principles to govern world trade in agriculture were considered. The United States suggested that 'there should be no intervention in imports and that as a minimum the basic GATT principles . . . should apply'. In response to a question as to whether the GATT waiver the United States had received in 1955 would 'have to be changed or eliminated', the United States' representative responded that the US would 'have to get rid of our Section 22 import quotas and other things like that', and that 'legislation would be required', noting that 'Our agriculture sector has been told these things are on the table and it is not going to be easy.'[35] Australia called for the '[p]rogressive reduction of the gap between administered internal prices and international market prices'. However, within the Negotiating Group on Agriculture, it was 'pointed out that where import restrictions were unavoidable it would be necessary to ensure that minimum access be provided'.[36]

During the June 1987 Venice G-7 Summit, while Japanese officials acknowledged the importance of concluding a Uruguay Round agreement that would lead to a reduction in agricultural subsidies, Japanese officials also stressed the importance of maintaining a certain degree of security in food supplies.[37] The Venice G-7 Economic Declaration reflected some of Japan's concerns, stating that the long-term objective was 'to allow market signals to influence the orientation of agricultural production, by way of a progressive and concerted reduction of agricultural support, as well as by all other appropriate means, giving consideration to social and other concerns, such as food security, environmental protection and overall employment'.[38] The Venice G-7 Economic Declaration reaffirmed the G-7 leaders' commitment to the agreement reached at the May 1987 OECD annual meeting,[39] at which the OECD had released statistics demonstrating that Japan had the highest agricultural producer subsidy equivalents for all available commodities, equivalent to '60 percent of the value of production, compared with 45 percent in the European Community and only 15 percent in the United States'.[40]

During the 6–7 July 1987 meeting of the Negotiating Group on Agriculture, the United States set out the first major negotiating proposal considered by the Negotiating Group. The United States called for a ten-year phase-out of import barriers, plus a similar phase-out of agricultural subsidies (including export subsidies) and an agreement concerning health and sanitary regulations. According to the proposal, 'measuring devices and an overall schedule of reductions should be agreed to for taking aggregate levels of support to zero over a 10-year period'. Also, 'special policy changes should be identified by each country to meet its overall commitment of scheduled support reductions, with these changes being agreed to by the other contracting parties'.[41] The EC's Director General of Agriculture, Guy Legras, expressed doubts about how realistic was the American proposal to reduce dramatically government intervention in agriculture.[42] The Japanese delegation reacted by stating that Japan would propose something more realistic, from the perspective of a country concerned with stabilizing its food supply. This would permit some import restrictions to continue under certain conditions.[43]

During the 26–27 October 1987 meeting of the Negotiating Group on Agriculture, the EC, the Cairns Group and Canada presented negotiating proposals, and several other participants provided indications of their initial positions.[44] The EC proposed a phased reduction of agricultural support policies in two stages: first, 'a series of short-term actions, based on existing policies (including emergency measures on cereals, sugar and dairy products)'; second, 'a concerted reduction in support, coupled with a readjustment of external protection in order to stabilize major world markets'.[45] The Cairns Group of agricultural exporting nations – which included Argentina, Australia, Brazil, Canada, Chile, Colombia, Hungary, Indonesia, Malaysia, New Zealand, the Philippines, Thailand and Uruguay – tabled a proposal that represented a compromise between the United States' and the EC's, but which was closer to the American one. The Cairns Group aimed to liberalize agricultural trade through a series of measures that would provide quick relief, as well as in the context of 'a long-term framework under which market access restrictions would be largely removed'.[46]

During late December 1987, Japan presented its first major proposal to the Negotiating Group on Agriculture. Japan's proposal 'stressed the need to establish stability for trade in agricultural products and to ensure food security for all countries'. The proposal 'pointed out that the national role of agriculture is not purely economic: it meets such social concerns as food security, environment protection and overall employment'. Japan 'considered that improved market access should be sought by reducing customs tariffs through a request-and-offer procedure, and also by allowing waivers from the general principles of quantitative restrictions'. Concerning basic foodstuffs, Japan 'called for new rules that would ensure stability of supply and allow the maintenance of domestic production' (GATT, *GATT Activities 1987*, 1988, p. 32; see also Breen, 1993, pp. 186–7).

Japan's proposal was discussed during the 15–17 February 1988 Negotiating Group on Agriculture meeting. Many countries' representatives were concerned about Japan's advocacy of ensuring food security through self-sufficiency, which they considered to be 'outdated, and a source of resource and trade distortions'. 'Several delegations said there was a lack of balance since [Japan's] proposals for action on the import side did not match those on the export side.'[47] The United States was unwilling to accept Japan's assertion that national food security had to be equated with ensuring certain levels of self-sufficiency in domestic production. The EC accused Japan of trying to evade its responsibility as a major importer to open up its markets to foreign agricultural products. An Australian delegate is reported to have stated that Japan was 'back-pedaling' from the May 1987 OECD annual meeting communiqué.[48]

The Mid-Term Review Agreement

During the 9–10 June 1988 meeting of the Uruguay Round Negotiating Group on Agriculture, while the EC further elaborated the short-term measures that it advocated,[49] the United States countered Japan's and other countries' concerns about food security by arguing that food security could be realized more effectively through a more liberalized trading system in which 'the elimination of restrictions on trade in food products would allow better supply, better allocation of resources and more stable prices'.[50]

During the June 1988 Toronto G-7 Summit meeting, Japan's Prime Minister Takeshita asked that other nations consider the particular circumstances of each country during the Uruguay Round agriculture negotiations, emphasizing Japan's low self-sufficiency rate, and noting that Japan was the world's number one net importer of agricultural products. Japanese officials stated that Japan was making efforts at agricultural reform and reducing price supports for domestic agricultural products, but noted that Japan did not intend to liberalize its rice market in the context of the Uruguay Round agriculture negotiations.[51] The Economic Declaration that resulted from the Toronto G-7 Summit meeting called for efforts in the Uruguay Round agriculture negotiations 'to adopt a framework approach, including short[-] as well as long-term elements which [would] promote the reform process'. Such a framework 'would be facilitated by a device for the measurement of support and protection'. The Declaration reflected some of Japan's concerns, noting that 'ways should be developed to take account of food security and social concerns'.[52]

During the 13–14 July 1988 meeting of the Uruguay Round Negotiating Group on Agriculture, the Cairns Group suggested that, as part of the Uruguay Round Mid-Term Review Agreement, the Ministers should agree that the negotiations concerning agriculture should concern, starting in 1989, among other issues, 'elaborating rules and disciplines for improving market access (elimination of tariff and non-tariff barriers, and likewise waivers and exceptional regimes for agriculture)'.[53] During discussion of the Cairns Group's proposal, a few delegates 'agreed that reliance on the international market alone was not sufficient to assure food security, and that some capability for self-production and the maintenance of some border measures were necessary'.[54]

During a 25–26 July 1988 meeting, the Uruguay Round Group of Negotiations on Goods reviewed progress in the ongoing Uruguay Round negotiations. Most participants 'underlined that improvement in Market Access remained a very important goal, and that the Agriculture negotiations played a central role' in the entire Uruguay Round negotiations.[55]

At the 12–13 September 1988 meeting of the Uruguay Round Negotiating Group on Agriculture, Japan discussed the possibilities for short-term, '"temporary"' measures that could be implemented after the Mid-Term Review Agreement was reached, 'while stressing that such measures would have to conform to the fundamental elements of the long-term framework which should be agreed upon at the same time'. Japan wanted each participant to be given the flexibility to select the concrete, short-term measures to be implemented. Japan suggested that Article xi of the GATT, which called for a general elimination of quantitative restrictions, be changed so that it permitted an exception for basic foodstuffs. Japan's intention, according to the GATT, 'was not to reinforce the scope of import restrictions but to recognize that a stable supply of basic foodstuffs was essential for every country from the point of view of food security, and especially for those with a low self-sufficiency rate'. Japanese negotiators argued that rice in Japan 'was an example where there was a "solid national consensus" that [an agricultural product] be supplied domestically even if at a higher cost'.[56]

During the meeting, 'One delegation questioned the contention that there were certain areas like rice for Japan where political requirements transcended economic logic', stating that such an assertion ran 'counter to the direction towards agricultural trade liberalization in which the Group should be moving'.[57] According to one source, chief US agriculture trade negotiator, Daniel Amstutz, indicated that the United States was prepared to wait and see how the liberalization of Japan's rice market might fit into the overall context of the negotiations.[58]

During the 11–14 October 1988 meeting of the Negotiating Group on Agriculture, Japan supplemented its December 1987 negotiating proposal. Japan's representative put forward 'the Japanese view that quantitative restrictions on basic foodstuffs should receive special treatment', and stated that short-term measures 'must be in line with and a part of the long-term objectives', and 'the selection of policies to be implemented should be left to the participants' individual decisions'.[59] South Korea 'emphasized the specific characteristics and national differences of agriculture, [agriculture's] non-economic aspects, and its fundamental importance in development'. South Korea's representative 'pointed out the weakness of Korean agricultural structures, and the important role that agriculture still played in the country though the Government was undertaking economic adjustment', arguing that the negotiations should 'allow developing countries such as Korea a sufficient period for structural adjustment, and some autonomy in the opening of their domestic markets'.[60]

During November 1988, the United States introduced the idea of comprehensive tariffication of agricultural products, proposing that the Ministers should agree in the Mid-Term Review Agreement 'to convert all non-tariff barriers, including variable levies and barriers maintained under waivers or other exceptions, into tariffs', and that participants should submit specific proposals for the rollback of non-tariff barriers and subsidies by January 1990. The United States suggested, in the Uruguay Round Mid-Term Review Agreement, that the Ministers should agree to a freeze on agricultural support and protection during 1989 and 1990.[61]

During late November 1988, the Chairman of the Negotiating Group on Agriculture proposed 'Points for Decision' for the Ministers attending the GATT December 1988 Montreal Mid-Term Review Ministerial Meeting; these included both 'long-term' and 'short-term' elements.[62] Concerning import access, the 'Points for Decision' document stated that 'all measures maintained under waivers, protocols of accession or other derogations and exceptions should be eliminated or brought under the strengthened GATT regime', and that 'conditions should be established governing the maintenance, elimination or removal in favor of tariffs, of quantitative or other non-tariff access restrictions and of measures not explicitly provided for in the General Agreement, including specification of access levels'.[63]

Just before the December 1988 Montreal Ministerial Meeting, President Reagan announced that, while other nations had called the United States' timetable 'unrealistic', US negotiators at the Montreal meeting would be 'flexible about timetables, so long as everyone agree[d] on nailing down an adjustment plan with specific dates for ending trade-distorting subsidies and market access barriers'. The president also announced that US negotiators at the upcoming meeting in Montreal would be prepared to 'talk about government stockpiles, land purchases, and other ways of dealing with what has been called food security'.[64]

At the beginning of the December 1988 Montreal GATT Ministerial Meeting, Japanese Foreign Minister Uno, in his opening speech, stressed food security, is quoted as emphasizing 'the need for ensuring the supply of fundamental foodstuffs, such as rice', while calling for '"a realistic approach, extending due consideration to the interests of both exporting and importing countries, in working toward the goals enunciated in the Punta del Este Declaration"'.[65]

During the December 1988 GATT Ministerial Meeting, the United States and the EC were unable to reach agreement concerning the agricultural provisions of the Mid-Term Review Agreement, particularly the provisions concerning agricultural subsidies. The United States sought to include language 'to *eliminate* trade-distortive support and protection', while the EC sought to include language in the Mid-Term Review Agreement which called for 'a *substantial reduction*' (GATT, *GATT Activities 1988*, 1989, p. 37). As a result of this impasse, most agricultural issues, including the market access provisions that might affect Japan's virtual ban on rice imports, were not discussed in much detail during the meeting.[66] Because an agreement was not reached concerning agriculture, as well as in three other areas, the Uruguay Round Mid-Term Review Agreement could not be finalized during the meeting. As a result, the Ministers decided 'that the nine subjects on which agreement had been secured should be put "on hold"' (GATT, *GATT Activities 1988*, 1989, p. 24), and that the Uruguay Round Trade Negotiations Committee should review the results achieved at a meeting of high-level officials during the first week of April 1989. Until that time, Trade Negotiations Committee Chairman, Arthur Dunkel, was to conduct 'high level consultations on the four items (Textiles and Clothing; Agriculture; Safeguards; and Trade Related Aspects of Intellectual Property Rights, Including Trade in Counterfeit Goods) which require[d] further consideration'. The Ministers reaffirmed their commitment to the objectives set out in the Punta del Este Declaration and 'resolved to maintain their efforts to achieve these objectives through the continuing work of the Negotiating Group on Agriculture'.[67]

At the closing session of the December 1988 GATT Ministerial Meeting, the United States stated that the Ministers had agreed not to settle 'for modest improvements in agricultural trade'.[68] Japan's Minister of International Trade and Industry, Hajime Tamura, regretted that the Mid-Term Review Agreement had not been finalized at the meeting. However, Tamura did recognize that, in the areas of services, tropical products, dispute resolution, and the strengthening of the GATT, very significant agreements had been obtained.[69]

Between December 1988 and April 1989, the United States, the EC, Japan and other countries resolved their remaining differences concerning the Uruguay Round Mid-Term Review Agreement.[70] The negotiations concerning the Mid-Term Review Agreement were completed during the early April 1989 Trade Negotiations Committee meeting, where agreement was reached concerning previously disputed provisions addressing agriculture, textiles and clothing, safeguards and trade-related aspects of intellectual property rights (TRIPS).

The sections concerning agriculture in the Uruguay Round Mid-Term Agreement stated that the Ministers agreed that the long-term objective was 'to provide for substantial progressive reductions in agricultural support and protection sustained over an agreed period of time, resulting in correcting and preventing restrictions and distortions in world agricultural markets'. The Agreement noted that, in realizing this long-term objective, 'the strengthened and more operationally effective GATT rules and disciplines, which would be equally applicable to all contracting parties, and the commitments to be negotiated, should encompass all measures affecting directly or indirectly import access and export competition'. The Mid-Term Review Agreement stated that the provisions concerning import access in the final agreement should concern 'quantitative and other non-tariff access restrictions, whether maintained under waivers, protocols of accession or other derogations and exceptions, and all measures not explicitly provided for in the General Agreement', and the matter of conversion of such measures into tariffs; as well as 'tariffs, including bindings'. The Agreement required the Ministers to provide by October 1989 specifics concerning their 'intention to reduce support and protection levels for 1990'. Participants were also 'invited to advance by December 1989' detailed proposals concerning long-term measures related to 'tariffication, decoupled income support, and other ways to adapt support and protection', as well as to submit, starting no later than December 1989, reports concerning their compliance with the agreement at six-month intervals. No later than the end of 1990, participants were to 'agree on the long-term reform programme and the period of time for its implementation'.[71]

The Japanese delegation was concerned that the Mid-Term Review Agreement did not deal sufficiently with food security, and did not take into consideration Japan's efforts to permit a certain domestic production level for staple foods.[72] But, after the Mid-Term Review Agreement was reached, as discussed in Chapter 4, officials of the Japanese Ministry of Agriculture, Forestry and Fisheries stated that, because the Mid-Term Review Agreement did not include a specific base year on which to calculate the conversion of import access measures into tariffs, the Agreement would not require Japan to make any concessions concerning access to its rice market in the short-term, and therefore would not affect Japan's food security in the short-term.[73]

During a late-July 1989 Trade Negotiations Committee meeting, three formal phases were established for the eighteen months remaining before the end of 1990, when the Uruguay Round negotiations were scheduled to be concluded. Until the end of 1989, national positions were to be presented to the negotiating groups. From January to July/August 1990, broad agreements were to be reached in each negotiating group. Between September 1990 and the final December 1990 Ministerial Meeting, final legal texts were to be prepared.[74]

The tabling of national positions

During July 1989, the GATT Secretariat released, at the request of the Negotiating Group on Agriculture, 'an overview of the existing framework of rules and disciplines relating to agriculture',[75] which included an extensive analysis of Article xi of the GATT that concerns the 'General Elimination of Quantitative Restrictions' on imports.

During the 10–12 July 1989 meeting of the Negotiating Group on Agriculture, the United States further elaborated its proposal concerning the comprehensive tariffication of agricultural products. The United States argued that the conversion of non-tariff import barriers to tariffs that had been included in the Mid-Term Review Agreement was 'the first, logical step towards liberalization of [agricultural market] access and was not intended to be an end to reform in this area: substantial progressive reductions in the tariffs thus established would follow'.[76] The United States' proposal discussed 'the methodology and mathematical formula that could be used in order to calculate the ad valorem equivalent of a non-tariff barrier'. The United States argued that tariffs were 'the least trade-distortive type of import barrier', and that converting non-tariff import barriers into tariffs 'would make world market prices more stable and predictable' and provide revenue. The proposal also noted that the reduction of tariffs 'was more readily negotiable' than the reduction of non-tariff barriers. The proposal argued that tariffication 'would permit the elimination not only of all restrictive measures but also of waivers and other exceptions under which restrictions on access were being maintained'.[77] During the meeting, it was noted that 'tariffication would pose problems for many participants if they had to commit themselves, by the end of the Round, to tariffication of all non-tariff barriers as well as eventual elimination of the tariffs thereafter'.[78]

During the 25–26 September 1989 meeting of the Negotiating Group on Agriculture, the EC's representative stated that 'the Community was not a defender of Article xi and did not want to extend its scope. However, certain elements in [Article xi], e.g. production controls and minimum access, should be maintained'.[79] Japan developed its ideas on 'the need to take full account of non-trade concerns, in particular food security and stable supplies of basic foodstuffs'. Japan considered it 'essential to maintain some level of domestic production, as food security cannot be ensured solely by the maintenance of potential production capacity, food stocks, bilateral agreements or diversification of suppliers'.[80]

During the 25–26 October 1989 meeting of the Negotiating Group on Agriculture, Switzerland argued 'that different situations must be treated differently', and that 'Article xi needed revision to impose operational disciplines and quantifiable minimum access commitments'.[81] The United States expanded on its plan for the comprehensive tariffication of agricultural products, suggesting that 'all non-tariff obstacles such as quotas, variable levies, voluntary restraint agreements, minimum import prices etc., would be converted into tariffs and bound'. 'All tariffs, including those resulting from such a conversion' would then 'be progressively reduced to zero or low levels over a ten-year period. All forms of derogation from existing GATT rules' should be eliminated, including Article xi:2(c), which authorized 'certain quantitative restrictions in the agricultural sector'. The United States also suggested that a safeguard mechanism would protect against import surges during the ten-year transition period.[82] Were the American proposal to be implemented, Japan would be required to replace its virtual ban on rice imports with a tariff that would make rice imported into Japan as expensive as rice produced in Japan. This tariff would then be lowered almost to zero over a ten-year period. Officials from Japan's Ministry of Agriculture, Forestry and Fisheries stated that the United States' proposal for tariffication would be difficult for Japan and other countries to implement.[83]

During the 27–28 November 1989 meeting of the Negotiating Group on Agriculture, the Cairns Group joined the United States in advocating the conversion of non-tariff measures on agricultural products to tariffs, and their progressive reduction, as well as the 'elimination of all provisions for exceptional treatment' and waivers in agriculture.[84] In contrast, Japan further elaborated conditions under which a country should be allowed to maintain import restrictions on agricultural products, such as it maintained on rice.[85]

South Korea proposed that the 'most reasonable alternative to ensure non-trade objectives, including food security, was to maintain potential agricultural production capabilities by introducing the concept of minimum market access or minimum rates of self-sufficiency levels'. South Korea aimed 'to clarify the rights and responsibilities of developing importing countries'. South Korea's representative argued that '[q]uantitative restrictions and internal support measures required for non-trade objectives should be permitted under strengthened GATT rules and disciplines'. Furthermore, '[i]n implementing long-term reforms, developing countries should be given longer time frames and flexibility in selecting policies and products'.[86]

During the 19–20 December 1989 meeting of the Negotiating Group on Agriculture, the EC indicated support, with some reservations, for the tariffication of agricultural products that the United States and the Cairns Group advocated.[87] But, the Nordic countries argued that 'the negotiations should define the exceptional cases under which quantitative restrictions would be allowed under a revised Article xi:2(c)',[88] and Austria emphasized that 'national policies aimed at securing food supply and other non-trade concerns had to be maintained'.[89]

The July 1990 Framework Agreement of de Zeeuw

During early 1990, the six ASEAN member countries and Colombia and Peru distanced themselves from Japan's advocacy of maintaining import restrictions for reasons of food security, and began to side with the Cairns Group's November 1989 call 'for preferential treatment for net importers that [were] developing nations'. The Cairns Group maintained that countries such as Japan and South Korea 'should not be permitted to preserve substantive farm supports due to the maturity of their economies'. Switzerland and the Nordic countries remained closer to supporting Japan's position concerning market access, because they, like Japan, were industrialized.[90]

As the potential impact of the Uruguay Round agriculture negotiations on the agricultural policies of various countries came to be discussed publicly, transnational coalitions formed among agricultural interest groups that might be affected by the negotiations. During February 1990, an international meeting of eleven international farm organizations in Geneva, organized by the Dairy Farmers of Canada, but which included representatives of the United States' National Farmers Union, as well as observers from Japan's Central Union of Agricultural Co-operatives, prepared a statement calling for the Uruguay Round negotiators to clarify Article xi of the GATT. According to the *International Trade Reporter*, a newsletter that tracks developments in US trade policy and international business, the organizations called for quantitative import restrictions to be 'based on historical imports of a given product in a previous representative period'. The organizations stated that 'In cases where historical [import] levels [were] nil or not representative of normal unimpeded trade, the initial quota should be based on a predetermined minimum access criteria. Once set, the initial quota would be subject to the same proportionality requirements as quotas based on historical imports.'[91]

During February 1990, the GATT Secretariat released a table summarizing the various negotiating proposals that addressed border measures submitted to the Negotiating Group on Agriculture by the United States, the Cairns Group, the EC, Japan, the Nordic countries, Switzerland, Austria, South Korea, Morocco and Bangladesh, as well as a joint proposal put forward by Brazil and Colombia, and a joint proposal of Egypt, Jamaica, Mexico, Morocco and Peru.[92] Proposals by India, Israel and Mexico were added to this table in a later revision.[93]

A few months later, during April 1990, the GATT Secretariat released responses to requests for clarification from the United States, the Cairns Group, the European Community and Japan concerning their stances on tariffication and import access.[94] In May 1990, the Secretariat also released responses to requests for clarifications from Japan, the Nordic countries, South Korea, Switzerland and Austria concerning their stances on non-trade concerns.[95]

During the 9 April 1990 meeting of the Uruguay Round Group of Negotiations on Goods, 'Many participants emphasized the central importance of Agriculture in the achievement of a satisfactory overall result' in the negotiations, and acknowledged that reaching an agreement on the broad outlines of a deal by July 1990 was 'essential'.[96]

During early July 1990, GATT Director-General Dunkel tried to persuade Japan's Agriculture Minister Tomio Yamamoto that it would be difficult for Japan to convince other countries to include food security in a Uruguay Round agriculture agreement.[97] But later in the same month, Yamamoto reiterated that Japan was not willing to alter its virtual ban on rice imports, noting that, while Japan had restricted the importing of 103 agricultural items some three decades before, there were currently restrictions on only seventeen items, and after the liberalization of orange and beef imports was completed during Spring 1991, there would be restrictions on only thirteen.[98]

During early July 1990, in an effort to produce a final negotiating text to meet the deadline set by the Trade Negotiations Committee, Negotiating Group on Agriculture Chairman de Zeeuw produced a Framework

Agreement for agriculture. This Agreement, as discussed in Chapters 3 and 4, contained many elements of the United States' proposals concerning comprehensive tariffication, but also provided 'a basis for taking account in the negotiation of non-trade concerns, special and differential treatment for developing countries, and the situation of net food-importing developing countries'. The Framework Agreement called for country lists to be submitted no later than 1 October 1990. These lists were supposed to be built on a number of parameters, including 'conversion of all border measures other than normal custom duties into tariff equivalents, irrespective of the level of existing tariffs'. The Framework Agreement also called for, 'in the case of absence of significant imports, establishment of a minimum level of access from 1991–92, also on the basis of tariff quotas at low or zero rate and representing at least [x] per cent of current domestic consumption of the product concerned'.[99]

The United States and the Cairns Group[100] expressed support in general for the Framework Agreement, but the EC expressed support for tariffication with some reservations. In contrast, Japan continued to assert that import restrictions for basic foodstuffs must be permitted, and that non-trade concerns should be accommodated outside the tariffication approach.

During early July 1990, the G-7 leaders issued the Houston G-7 Summit Economic Declaration, which commended to the GATT member countries' negotiators de Zeeuw's Framework Agreement 'as a means to intensify the negotiations'.[101] The Summit's Economic Declaration reflected concerns about food security raised by Japan's Prime Minister Kaifu[102], incorporating wording that the Uruguay Round negotiations should 'be conducted in a framework that includes a common instrument of measurement, provides for commitments to be made in an equitable way among all countries, and takes into account concerns about food security'.[103]

During a late-July 1990 meeting of the Uruguay Round Trade Negotiations Committee, GATT Director-General Dunkel echoed the Houston G-7 Summit Economic Declaration, describing the de Zeeuw text as 'a means to intensify the negotiations', and called on countries to present their country lists of concessions by 1 October 1990 and their offers by 15 October 1990.[104]

The December 1990 Ministerial Meeting

At the 27–29 August 1990 meeting of the Negotiating Group on Agriculture, Japan submitted a paper that proposed the review of, and additions to, Articles XI, XX, and XXI of the GATT in order to establish rules for basic foodstuffs.[105] In late September 1990, Japan submitted a list of agricultural concessions, as well as an offer list, to the GATT. Japan's Ministry of Agriculture, Forestry and Fisheries announced plans to reduce agricultural subsidies by 30 per cent over 10 years,[106] but refused to agree to comprehensive tariffication, maintaining Japan's right to virtually ban rice imports.[107]

By 4 October 1990, only Argentina, Australia, Austria, Canada, Finland, Japan, New Zealand, Singapore, Sweden and Switzerland had submitted lists of concessions in accordance with GATT Director-General Dunkel's 1 October 1990 deadline.[108]

During early October 1990 bilateral meetings, the United States complained that Japan had not fully accepted the comprehensive tariffication of agricultural products advocated by the United States in the Uruguay Round agriculture negotiations.[109] Two weeks later, the United States put forward its offer, calling for '75 percent reductions each in internal supports and market access barriers'. The United States also proposed that, along with agreeing to comprehensive tariffication, countries should agree to a certain level of minimum access to agricultural markets. Countries that currently had import restrictions on agricultural products 'should start to import at least 3 percent of their annual domestic consumption of any import-restricted item with no or few tariffs'. Tariffs 'on the remaining 97 percent should be lowered in 10 years to 50 percent'. Furthermore, this '3 percent free-import ceiling should be gradually raised to 5.25 percent in 10 years, after which all imports of the item should be conducted with tariffs under 50 percent'. This proposal meant that Japan, which consumed 10 million tons of rice annually, would initially be required to import at least 3 per cent of its annual domestic consumption (or 300,000 tons) of rice.[110]

By 29 October 1990, twenty-seven countries, including the United States and Japan, had submitted country lists, and ten countries, including the United States and Japan, as well as the Cairns Group (apart from Canada) had submitted offer lists.[111] By early November 1990, the EC had still not submitted an offer list.

At the beginning of the December 1990 Brussels GATT Ministerial Meeting during which the Uruguay Round negotiations were scheduled to be concluded, GATT Director-General Dunkel, as Chairman of the Trade Negotiating Committee, released a 400-page draft of the Uruguay Round agreements which included a section on agriculture. This incorporated not only de Zeeuw's July 1990 Framework

Agreement, but also Dunkel's 'Principal Issues for Decision' and a 'Survey of Offers' concerning agriculture. Dunkel noted that most participants had included tariffication in their offers, but India and Japan had not, and a number of countries had limited tariffication to certain products. Japan, along with Austria, Canada, Finland, Iceland, South Korea, Norway and Switzerland, had called for the maintenance of quantitative import restrictions under defined conditions.[112]

During the December 1990 Ministerial Meeting, a last-minute effort was made to reach an agreement concerning agriculture, using a compromise proposal tabled by Swedish Minister of Agriculture, Mats Hellstrom, who was appointed Chairman of the Negotiating Group on Agriculture at the Ministerial level for the meeting.[113] One of the central points of disagreement was minimum access to agricultural markets. The Hellstrom proposal stated that 'in the case of absence of significant levels of imports', a minimum level of access from 1991–2 would be established, which represented 'at least 5% of current domestic consumption of the product concerned'.[114] The EC, pushed by France and Ireland, initially found itself isolated in rejecting the Hellstrom proposal, but then Japan and South Korea, who were concerned about the minimum access provisions, also joined in rejecting it.[115] Yamamoto, Japan's Minister of Agriculture, Forestry and Fisheries, complained that the Hellstrom proposal did not reflect adequately concerns about food security that Japan had been emphasizing throughout the negotiations.[116]

Unable to reach an agreement concerning agriculture and therefore unable to conclude the Uruguay Round negotiations on schedule, the Ministers present at the December 1990 Ministerial Meeting agreed that Dunkel, as Trade Negotiations Committee Chairman, should engage in '"intensive consultations"' until the beginning of 1991 with the objective of reaching agreements in all of the areas where differences remained (GATT, *GATT Activities 1990*, 1991, p. 27).

During early January 1991, Japan's Chief Cabinet Secretary, Misoji Sakamoto, announced that when a compromise between the United States and the EC was reached concerning agriculture, the problem of the opening of the Japanese rice market would become the next focus of the agriculture negotiations.[117]

During an early January 1991 visit to the United States, Japan's Foreign Minister, Tarō Nakayama, informed USTR Hills that Japan was willing to 'negotiate on the three key elements of global farm reform – domestic support, border measures, and export subsidies'. But Nakayama repeated the Japanese government's rejection of the Hellstrom proposal, stating that the proposal's call for a minimum access of five per cent of domestic consumption did not address the issues Japan had raised during the negotiations, such as non-trade concerns (for example, food security, or the need for Japan to maintain certain production adjustment programs).[118]

During late February 1991, while suggesting a tentative agenda for the ongoing negotiations, Dunkel, in his role as Chairman of the Trade Negotiations Committee, emphasized that a number of technical issues concerning market access remained to be addressed, including 'the modality and scope of tariffication', and 'the scope and modalities of implementation of a minimum access commitment'.[119]

During April 1991, the Uruguay Round negotiations were reorganized into six new negotiating groups, covering market access, textiles and clothing, agriculture, rule-making, trade-related aspects of intellectual property rights (TRIPS), and institutions. Dunkel was appointed Chairman of the Agriculture Group, and Germain Denis of Canada was appointed chair of the Market Access Group.[120] Dunkel informed the Trade Negotiations Committee that the new structure was 'flexible and result-oriented' and was 'designed to encourage substantive negotiations and to help achieve the political breakthroughs with the minimum delay'.[121]

The December 1991 Draft Final Act

During an early November 1991 assessment of the negotiations, Dunkel, the Chairman of the Uruguay Round Trade Negotiations Committee stated that, in the agriculture negotiations, tariffication was 'emerging as the fundamental pillar of the reform process', and decisions were needed on 'What would be the conditions for, and the size of, the minimum access commitments and on what basis would they be undertaken? Would minimum access opportunities expand over time?'[122] Dunkel also stated that the market access negotiations needed 'an early agreement on modalities, including tariffication, for negotiating market access as part of the reform programme being negotiated in agriculture'.[123]

During a November 1991 visit to Japan, USTR Hills argued that if Japan did not alter its stance concerning its virtual ban on rice imports, it would stand alone with South Korea in the Uruguay Round negotiations in rejecting the comprehensive tariffication of agricultural products.[124] Japan's Minister of Agriculture, Forestry and Fisheries, Tanabu, responded that Japan could not accept the comprehensive tariffication of agricultural products, because domestic opposition in Japan to the liberalization of the Japanese rice market was too strong.[125]

During late December 1991, Dunkel released the Draft Final Act Embodying the Results of the Uruguay Round of Multilateral Trade Negotiations. On the release of the Draft Final Act, Dunkel, as Trade Negotiations Committee Chairman, stated that while 'the entire body of agreements must be reviewed for legal conformity and internal consistency', such a process should not lead to 'substantive changes in the balance of rights and obligations established in the agreements'. Dunkel gave the contracting parties until 13 January 1992 to develop their positions.[126]

Despite Japan's and other countries' concerns, the 'Text on Agriculture' included in the December 1991 Draft Final Act stated that tariffication should occur for 'all border measures other than ordinary customs duties', including quantitative import restrictions.[127] The Draft Final Act stated that 'Ordinary customs duties, including those resulting from tariffication, shall be reduced, from the year 1993 to the year 1999, on a simple average basis by 36 per cent with a minimum rate of reduction of 15 per cent for each tariff line.' Concerning minimum access to agricultural markets, the Draft Final Act stated that 'Where there are no significant imports, minimum access opportunities shall be established.' These minimum access opportunities 'shall represent in the first year of the [six- to seven-year] implementation period not less than 3 per cent of corresponding domestic consumption in the base period . . . and shall be expanded to reach 5 per cent of that base figure by the end of the implementation period'.[128]

Shortly after the Draft Final Act was released, Tanabu, Japan's Minister of Agriculture, Forestry and Fisheries stated that Japan could not accept the provisions concerning tariffication without exception that were included in the Draft Final Act.[129]

But during the 13 January 1992 meeting of the Uruguay Round Trade Negotiations Committee, 'there was an overwhelming view that the Draft Final Act provided the best, indeed the only, basis on which to secure a speedy conclusion of the Round'.[130] While the United States stated that the Draft Final Act provided 'the necessary elements for finalizing the Round', Japan stated that 'the concept of comprehensive tariffication posed extreme difficulties for Japan'.[131] All countries agreed to participate in a new four-track negotiating strategy. One track was devoted to market-access negotiations at the bilateral, plurilateral and multilateral levels and focused on reaching 'specific commitments on internal support and export competition in agriculture'. The other three tracks concerned services, legal drafting and adjusting the Draft Final Act to any changes made.[132]

During late January 1992, Japan's Prime Minister Miyazawa, in response to a question concerning when Japan might have to decide whether to alter Japan's virtual ban on rice imports in order to reach agreement in the Uruguay Round negotiations, stated that Japan would not let the Uruguay Round fail, but would watch closely the progress of the negotiations concerning agriculture that were occurring between the United States and European Community.[133]

By the end of March 1992, in accord with the Draft Final Act, the United States and Japan submitted schedules of agricultural concessions to the GATT. But Japan left the liberalization of import restrictions on rice and dairy products off its list of concessions, and Japan's Ministry of Agriculture, Forestry and Fisheries stated that it would try to prevent Japan's rice market from being affected by the comprehensive tariffication of agricultural products that was included in the Draft Final Act.[134]

By April 1992, according to the GATT, twenty-three participants had presented schedules of agricultural concessions to the Negotiating Group on Agriculture, but these were not 'wholly in line with the provisions in the Draft Final Act' (GATT, *GATT Activities 1992*, 1993, p. 16). Dunkel, Chairman of the Negotiating Group on Agriculture, is reported to have called it '"progress"' that 'Japan now [had] its position on agricultural concessions on the table'. Dunkel noted that '"Three months ago we didn't know what they wanted, but now we know what the limits are."'[135] As for the Negotiating Group on Market Access, while all the participants had made a commitment 'to present draft final line-by-line market-access schedules covering all products by 31 March, the Secretariat had only received 37 submissions with a further 14 participants having indicated an intention to submit comprehensive draft schedules in the near future' (GATT, *GATT Activities 1992*, 1993, p.16).

During the rest of 1992, negotiations concerning agriculture continued between the United States and the EC, particularly about the export subsidy provisions in the Draft Final Act (Breen, 1993, pp. 218–27). Japan continued to wait, and at times to pressure, for a breakthrough in the US–EC negotiations. During the November 1992 Trade Negotiations Committee meeting, Japan 'strongly urged, as it had done in the past, that the multilateral negotiating process in Geneva be re-started as soon as possible and that the difficult questions that participants other than the European Community and the United

States might have in respect of the proposed Draft Final Act also be duly dealt with in that process'.[136]

After France approved the Maastricht Treaty in September 1992, and after Dunkel, supported by the Uruguay Round Trade Negotiations Committee that he chaired, appealed to the United States and the EC in November 1992 to reach an agreement concerning agriculture that would move the Uruguay Round negotiations forward,[137] the United States and the EC reached the Blair House Accord. The November 1992 Blair House Accord involved 'jointly-presented proposals to amend elements' of the agricultural provisions of the Draft Final Act, excluding certain forms of domestic support from the commitments included in the Act, lessening commitments to reduce export subsidies, and clarifying issues related to the resolution of the US–EC oilseeds dispute (GATT, *GATT Activities 1992*, 1993, p. 18).

The conclusion of the negotiations

Towards the end of 1992, forty-four countries 'had submitted draft schedules to the market access group', and efforts were made to agree on specific adjustments to the Draft Final Act so that the Uruguay Round negotiations could be concluded by the 2 March 1993 deadline established by the United States' existing fast-track negotiating authority (GATT, *GATT Activities 1992*, 1993, pp. 16, 19). Because of the November 1992 Blair House Accord, the United States and the EC were not able to present their revised schedules of agricultural commitments until late December 1992. Progress in the negotiations was delayed as a result of the November 1992 United States' presidential election and the need for the incoming Clinton Administration to receive Congressional approval for an extension of its negotiating authority, which it eventually received in June 1993. Japan and South Korea continued to wrestle with how to allay farmers' fears about the impact of the tariffication of rice and the lessening of restrictions on the import of rice required by the Draft Final Act. Meanwhile, farmers in Switzerland and Canada were calling for their governments 'to maintain quantitative restrictions on other farm products' (GATT, *GATT Activities 1993*, 1994, pp. 16–17).

During the December 1992 Trade Negotiations Committee meeting, the United States called for 'a meaningful negotiation of the agricultural market access package'. South Korea expressed disappointment concerning the 'common failure to work out a solution to other important pending issues, including tariffication in agriculture'.[138]

A month later, during the January 1993 Uruguay Round Trade Negoti-ations Committee meeting, Canada 'continued to argue its position on Article XI with regard to exceptions to comprehensive tariffication'.[139]

During an early July 1993 quadrilateral meeting, the United States, Japan, Canada and the EC reached a general agreement concerning market access, although the agreement contained few specifics con-cerning access to agricultural markets.[140] The July 1993 Tokyo G-7 Summit Final Declaration welcomed the market-access package that had been worked out at the July 1993 quadrilateral meeting,[141] stating that the Uruguay Round negotiations should be concluded by the end of 1993, and should include an agricultural market access package.[142]

During late July 1993, newly appointed GATT Director-General Peter Sutherland, who replaced Dunkel as Chairman of the Trade Negoti-ations Committee, suggested a calendar by which to work out an agri-cultural market access agreement. This calendar was not derestricted until 1997. Sutherland called for a 'series of bilateral negotiations on agricultural market access' to be held 'in Geneva beginning 30 August through the week of 13 September' leading to a substantive stocktak-ing meeting on 15 October 1993, followed by the 'submission of agreed changes and revisions to the Draft Schedules of Concessions by 15 November', and a final market-access agreement by 15 December 1993. The United States said that 'the autumn programme outlined by the Chairman was very realistic and useful and could in fact work to move participants towards a successful outcome'. Japan 'entirely agreed with the Chairman[']s proposed work programme' and 'stood ready to take part in the process ahead in order to bring about a suc-cessful conclusion of the Round by the year's end'.[143]

During September 1993, Sutherland made reassuring statements con-cerning the market access agreement under discussion, noting that '[i]mport prohibitions will have to be replaced by a tariff-based system, but the new system may provide, initially, an equivalent level of protec-tion to the existing systems'. Sutherland stated that the level of pro-tection would 'be subject to a gradual reduction over a six-year period – by no means a complete opening of borders to foreign products'.[144]

During late September 1993, USTR Kantor is reported to have stated that the fact that Japan, as a result of a poor rice harvest, had recently announced that it would purchase rice from abroad '"create[d] a more positive atmosphere"' for the market access negotiations.[145] Before departing for Japan in October 1993, US Agriculture Secretary Espy reportedly stated '"We've got to have a market access formula before we will sign a GATT agreement, and that should include Japan, and that should include questions of rice"'.[146] Secretary Espy encouraged Japanese officials to open the Japanese rice market to imports in order to conclude the Uruguay Round negotiations, but Japan's Agriculture Minister, Eijirō Hata, responded that Japan's recent imports of rice were an emergency measure, and did not signal any change in Japan's negotiating stance at the Uruguay Round.[147]

On 15 October 1993, the *Asahi Shimbun*, a Japanese newspaper, reported from Geneva that the United States and Japan were discussing a compromise plan that would amend the December 1991 Draft Final Act. Japanese negotiators sought to maintain Japan's virtual ban on rice imports, while the United States wanted rice imports to be allowed, even if they were subject to high tariffs. The plan under discussion would require Japan to establish an initial minimum access of 4.5 per cent of its domestic consumption of rice, totaling about 450,000 tons, which would rise over a six-year period to 7.5 per cent.[148] It was also reported that Japan had made a definitive decision to allow the tariffication of about 20 other agricultural products, including wheat, barley and dairy products.[149] But, Japan's proposals to the GATT did not address directly Japan's virtual ban on rice imports, even though GATT officials stated that Japan's virtual ban on rice imports remained on the agenda.[150]

During early November 1993, the *International Trade Reporter* noted that US Agriculture Secretary Espy had recently announced that Japan would import 'between 300,000 and 500,000 metric tons of rice from California and other states' as part of its emergency measures resulting from a poor rice harvest.[151]

During the 19 November 1993 Uruguay Round Trade Negotiations Committee meeting, Chairman Sutherland highlighted agriculture and four other issues that remained to be addressed, stating that 'comprehensive tariffication and the provision for current and minimum access opportunities, remained key to the achievement of a satisfactory global market-access package'. By that time, eighty-three countries had presented market access offers.[152]

After the November 1993 APEC meeting in Seattle, during which agreement was reached concerning a number of conditional market access offers that were related to the July 1993 quadrilateral market access agreement (GATT, *GATT Activities 1993*, 1994, p. 21), Japanese Prime Minister Hosokawa called for the successful conclusion of the Uruguay Round negotiations, stating that the rice issue would have to be resolved alongside all the difficult issues that other countries were dealing with.[153]

During the 26 November 1993 Uruguay Round Trade Negotiations Committee meeting, South Korea announced that it 'was ready to reduce the scope of tariffication exceptions to the very minimum', but 'maintained the position that exceptionally sensitive items, including basic foodstuffs, should not be subject to tariffication'.[154]

Towards the end of November 1993, Sutherland, in an effort to conclude a Uruguay Round agreement by the 15 December 1993 deadline established by the June 1993 US fast-track negotiating authority legislation, called frequent informal meetings of the Heads of Delegations 'to "gavel through" the less controversial texts' of the agreement, 'as well as to discuss other, more difficult, issues', among which were the demands of some countries, including Japan, 'that certain agricultural commodities – rice for instance – should not be subject to tariffication' (GATT, *GATT Activities 1993*, 1994, p. 23).

During early December 1993, the agricultural market access compromise plan under discussion, which addressed some of Japan's concerns, became public. The proposal, put forward by Germain Denis, Chairman of the Negotiating Group on Market Access, would, among other things:

- 'impose a six-year moratorium on the conversion of non-tariff trade barriers into tariffs for farm products whose imports comprise[d] less than 3 percent of domestic consumption, whose exports [were] not subsidized, and whose production [was] regulated domestically';
- 'give special treatment for rice on the condition that Japan [allow] minimum access for imports of between 4 percent and 8 percent of domestic rice consumption to imports during the six-year period'; and
- 'call[ed] for negotiations one year before the end of the six-year period to determine if special treatment [would] be continued after the moratorium ends'.[155]

On 7 December 1993, the day after USTR Kantor and EC External Economic Affairs Commissioner Sir Leon Brittan agreed to some small changes to the Blair House Accord,[156] and the day after South Korea announced that it would partially open its rice market (Breen, 1999, pp. 19–21), Prime Minister Hosokawa stated that Japan would shortly make a formal announcement concerning the partial liberalization of Japan's rice market.[157]

During the early morning of 14 December 1993, Kantor and Brittan finally concluded their negotiations. On the same day, Prime Minister Hosokawa announced that, in order to conclude the Uruguay Round negotiations, Japan would partially liberalize its rice market, in accordance with the minimum access provisions included in the Uruguay Round Agreement on Agriculture. During the following twenty-four hours, the sections of the Final Act Embodying the Results of the Uruguay Round that had yet to be approved 'were adopted informally in the Heads of Delegations meetings and last minute negotiations on market access took place among delegations' (GATT, *GATT Activities 1993*, 1994, p. 24).

On 15 December 1993, the delegates to the GATT approved the 400-page Final Act Embodying the Results of the Uruguay Round, which included the Uruguay Round Agreement on Agriculture. '[Ninety-five] participants had submitted draft Schedules of concessions covering both agricultural and non-agricultural products. A large number of participants had also confirmed that their offers were now final', and that the concessions they had exchanged would be duly reflected in the final schedules they would present to the GATT.[158]

The signing of the agreements and passage of implementing legislation

Between 15 December 1993 and the signing of the Uruguay Round agreements in April 1994, the Final Act Embodying the Results of the Uruguay Round 'was legally verified, agreed and circulated to all participants'. The schedules of the market access commitments were 'multilaterally verified for attachment to the Marrakesh Protocol'. On 15 April 1994, in Marrakesh, Morocco, the United States, Japan, and more than 100 other nations, signed the Uruguay Round agreements, including the Agreement on Agriculture. A GATT press release noted that 'The Round had lasted approximately 2,700 days and produced a package of results running to about 27,000 pages which reportedly weighed in at 175 kilos' (GATT, *GATT Activities 1994–1995*, 1996, pp. 15–16).

During early December 1994, GATT Director-General Peter Sutherland announced that the US Congress's passage of the Uruguay Round Agreements Act had cleared away one of the last obstacles to enable the World Trade Organization to come into existence on 1 January 1995. News on 1 December 1994 of the Senate's passing of implementing legislation 'arrived at GATT headquarters almost simultaneously with news that the penultimate hurdle for Japanese ratification had been cleared by a Diet committee in Tokyo'.[159] According to the GATT, '[w]hen the WTO entered into force on 1 January 1995, 76 governments were eligible to become members on the first day. Another 50 governments were at various stages of completing their domestic ratification procedures' (GATT, *GATT Activities 1994–1995*, 1996, p. 5).

Domestic politics and international relations in US–Japan trade policymaking during the negotiations

Based on the preceding discussion and the discussion in Chapters 3 and 4, the validity can be assessed of the three hypotheses set out at the end of Chapter 2, concerning the extent to which domestic politics affects the agreement the United States and Japan reach in a trade negotiation.

Concerning the first hypothesis, as discussed in Chapter 3, there was widespread domestic support in the United States for settling on and maintaining a free-trade-oriented negotiating stance in the Uruguay Round agriculture negotiations. This widespread domestic support made it easier for the United States to seek the cooperation of other countries in reaching an agreement concerning agricultural trade liberalization. As Theodore Cohn notes, American proposals for agricultural reform in the Uruguay Round 'were more broad ranging', and threats by the United States 'to link the agricultural talks with the outcome of the entire round were more serious' than in previous GATT rounds (Cohn, 1993a, p. 36). But the United States encountered more difficulties in concluding the Uruguay Round than in earlier GATT Rounds, partly because the Uruguay Round negotiations included many areas that previous GATT rounds had not covered so extensively – agriculture, intellectual property rights, services and investment, as well as the GATT rules themselves (Janow, 1994, p. 58). Also, the United States' relative decline within the international system made it more difficult for the United States to pressure the EC and Japan into reaching an agreement.

It is possible that a different presentation by the United States of its arguments concerning comprehensive tariffication and minimum access, a presentation which included a more complex market access scheme earlier in the negotiations, might have encouraged Japan to cooperate earlier in helping to conclude the Uruguay Round agriculture negotiations. But, whether a more complex market access scheme such as this would have received adequate domestic support in other GATT member countries is open to question.

In contrast, Japan, while generally agreeing to the agricultural export subsidy provisions under discussion, adopted a protectionist negotiating stance and opposed a mutual agreement that included the comprehensive tariffication of agricultural products and minimum access to agricultural markets. Japan's negotiators emphasized the strength of Japanese domestic opposition to rice liberalization, which was discussed in Chapter 4, in order to convince the United States and other countries that there were definite limits within which Japan would be able to alter its virtual ban on rice imports in order to reach a Uruguay Round agreement.

After a GATT dispute panel ruled in October 1987 that Japanese import restrictions on twelve agricultural products were in violation of Article XI of the GATT, many Japanese leaders believed that a similar GATT panel could be established to determine that Japan's virtual ban on rice imports was also in violation of Article XI of the GATT (Matsushita, 1993b, pp. 282–6). Therefore, Japan was adamant about not negotiating rice liberalization at the bilateral level, and committed to discussing the possible liberalization of the rice market at the multilateral level in the context of the Uruguay Round negotiations, where liberalization, if it had to occur, might be minimized and delayed for a number of years. Discussing Japan's virtual ban on rice imports in the Uruguay Round negotiations added more parties to the negotiations, which was of some use to Japan, because some countries, such as South Korea, shared similar concerns about the comprehensive tariffication of agricultural products and the minimum access provisions under discussion. Japan delayed making a decision concerning rice liberalization until the last day of the negotiations, bargaining energetically to be permitted to postpone the implementation of the agricultural concessions that it agreed to for a number of years, displaying one of the tendencies of Japanese negotiators highlighted by Michael Blaker, that Japanese negotiators tend to 'stick doggedly to original positions' (Blaker, 1977a, p. 100 see also Blaker, 1999, pp. 54–60).

As a result of domestic support in Japan for a protectionist negotiating stance, it was difficult for the country to cooperate fully with the United States and other countries in reaching an agreement concerning agricultural trade liberalization in the Uruguay Round negotiations. Because Japan was the only G-7 country that was very dependent on agricultural imports, it had difficulty gaining the sympathy of the other major industrialized countries, who were mostly farm exporters, and therefore could not relate much to Japan's lack of self-sufficiency in food. In the late 1980s and early 1990s, Japan's increased importance in the international system as a result of having ascended to the position of the second-largest economy in the world, gave the country greater leverage in the Uruguay Round agriculture negotiations. Japan was slow in attempting to form a bloc of countries (which eventually included South Korea) that would oppose the provisions concerning comprehensive tariffication and minimum access included in de Zeeuw's July 1990 Framework Agreement and in the December 1991 Draft Final Act. When Japan attempted to form such a bloc, the Cairns Group co-opted the developing countries, and Japan was disappointed at the extent to which Japan's advocacy of food security was not that popular among other GATT member countries. The Japanese press often portrayed Japan as a victim of an international trade regime that it had little involvement in creating. The 'compromise of embedded liberalism' that was settled on in 1947 at the GATT preparatory meetings was established long before Japan became a member of the GATT. The incorporation into the compromise of the comprehensive tariffication of agriculture was portrayed by Japanese negotiators as having potentially disastrous implications for Japan's rice market and for the country's farmers.

The Japanese government maintained its protectionist negotiating stance, and did not compromise on the liberalization of the rice market until the end of the negotiations, even though, as discussed in Chapter 4, the domestic coalition in Japan supporting such a stance eroded away gradually during the negotiations.

As for the validity of the second hypothesis, because there was little division in the American government concerning the negotiating stance that the United States should take in the Uruguay Round agricultural market access negotiations, the US Congress, while monitoring the negotiations actively, did not have a great deal of influence on the negotiations. There was not much domestic pressure for the United States government to alter the United States' negotiating stance substantially during the negotiations. Throughout the negotiations, the Rice Millers' Association continually pressured the Executive Branch and Congress in order to obtain the assistance of the United States government in opening up the Japanese rice market. When USTR Yeutter rejected the Rice Millers' Association's September 1986 Section 301 petition, the US government made the decision to negotiate the liberalization of Japan's rice market at the multilateral level in the Uruguay Round agriculture negotiations rather than at the bilateral level. While the United States' advocacy of comprehensive tariffication and minimum access to agricultural markets was a concern to

sugar, dairy, peanut and other agricultural producers who could potentially be adversely affected, the concessions involved were considered
to be relatively minor in relation to the benefits that might result from
a Uruguay Round agriculture agreement.

In contrast, there was much more division within the Japanese government concerning the negotiating stance Japan should take during
the Uruguay Round agriculture negotiations. The fact that such a division existed increased the influence of the Diet, whose members
became advocates on both sides of the debate – and some of whom
eventually supported the compromise plan worked out at the very end
of the negotiations which included an exception for Japan's rice
market.

Within the Japanese bureaucracy, bureaucrats from the Ministry of
Agriculture, Forestry and Fisheries were particularly concerned with
the asymmetry between Japan's agricultural self-sufficiency and that of
the United States and other countries, and sought to protect the inde-
pendence of Japanese agriculture. In contrast, the more internationally-oriented ministries, such as the Ministry of International Trade
and Industry, and the Ministry of Foreign Affairs, were concerned with
reducing US–Japan trade friction, and hoped to realize some gain by
cooperating with the United States and other countries on agricultural
import liberalization. However, these ministries tended to state that in
the end, rice policy and the decision concerning rice liberalization
were the responsibility of the Ministry of Agriculture, Forestry and
Fisheries.

There was relatively widespread opposition within the Diet to a
Uruguay Round agreement that would require some form of liberalization of Japan's rice market. But several prominent Diet members, including members of the LDP and Komeito, who were concerned with
trade friction between the United States and Japan, as well as with
capturing the urban and business vote, came to advocate some form of
liberalization of Japan's rice market in order to promote the conclusion
of the Uruguay Round negotiations. In contrast, members of the Japan
Socialist Party and other Diet members opposed any form of liberalization of Japan's virtual ban on rice imports, in order to secure the support of Japan's farmers.

These divisions within the Japanese bureaucracy and the Diet, as
discussed in Chapter 4, led to an active domestic debate concerning a
variety of policy options Japan could adopt in order to conclude the
Uruguay Round negotiations. These divisions increased the influence
of the Diet, which made it more likely that the Japanese government

would be able to alter its negotiating stance to enable the conclusion of the Uruguay Round negotiations.

As for the validity of the third hypothesis, because a substantial number of informed domestic groups in both the United States and Japan endorsed the negotiating stances of their governments, the US Congress and the Japanese Diet were more likely to endorse the results of the negotiations, and the chances of the two countries reaching a cooperative agreement concerning comprehensive tariffication and minimum access to agricultural markets increased.

Since a substantial number of informed domestic groups endorsed the United States' negotiating stance, the US Congress was likely to endorse a Uruguay Round agreement including an agreement that covered provisions concerning the comprehensive tariffication of agricultural products and minimum access to agricultural markets. As a result, the United States was able more easily to convince other countries to agree to the inclusion of such provisions in the Uruguay Round Agreement on Agriculture. As discussed in Chapter 3, the United States' proposals to convert import restrictions on agricultural products to tariffs, and to reduce them, as well as to provide minimum access to agricultural products, were believed to require small increases in United States' imports of a few agricultural commodities, which had relatively few domestic political ramifications in the United States (United States Agriculture Technical Advisory Committee, 1994). There was little domestic opposition to the negotiating stance of the United States, which increased the likelihood that the US Congress would endorse the results of the negotiations, and that the United States, Japan and other countries would reach an agreement.

In contrast, Japanese negotiators were under much greater pressure from various informed domestic groups who supported and opposed the Japanese government's negotiating stance at the Uruguay Round agriculture negotiations. Japan's heavily subsidized farmers, led by the Central Union of Agricultural Co-operatives, called for the continuation of Japan's virtual ban on rice imports. The older populace and Japan's agricultural sector were more concerned with Japan's vulnerability in terms of food supply, while the younger generation and urban consumers were more confident in Japan's economic strength, and less worried about the country's ability to feed itself. Many Japanese businessmen who were concerned with the overall efficiency of the Japanese economy and reducing the cost of Japan's food control system advocated agreeing to some form of liberalization of Japan's virtual ban on rice imports in order to conclude the Uruguay Round agriculture negotiations. However, such a change in agricultural policy posed a threat to long-subsidized farmers and Japan's agricultural co-operatives.

Because the Diet had issued several resolutions during the 1980s that expressed opposition to the liberalization of Japan's rice market, many were concerned that the Diet might not ratify a Uruguay Round agreement that required some form of liberalization of this market. Successful ratification was also called into question when the LDP could not gather greater support for agricultural reform without creating more fundamental divisions within the party, and this further weakened the LDP's longstanding dominance within the Diet. As Japan's negotiators often emphasized, were the United States' negotiating proposals to be accepted in the final Uruguay Round agreement, Japan might have to implement agricultural policies that would so weaken Diet members' links to their rural constituents that the Japanese Diet might reject complelety the Uruguay Round agreements. Such concerns reduced the United States' capacity to convince Japan and other countries to cooperate in reaching an agreement that included comprehensive tariffication and minimum access provisions.

As support among informed interest groups for Japan's initial negotiating stance gradually eroded during the course of the Uruguay Round negotiations, it became possible for the Japanese government to alter its original negotiating stance and to agree to some form of liberalization of the Japanese rice market in order to conclude the Uruguay Round negotiations with some reasonable assurance that the Diet would ratify the resulting agreement. During the negotiations, the pro-rice-market-liberalization coalition within Japan strengthened, and the final concessions Japan made were not that significant, and had been debated for long enough, that even those in Japan's anti-rice-market-liberalization coalition could see the importance of making some form of compromise in order to conclude the negotiations.

6
Conclusion

The approach to analyzing trade policymaking developed in this book strives to improve on Putnam's two-level game approach and to define more systematically, with reference to the existing literature, the nature of and relationship between domestic politics and international relations in one particular area – trade policymaking.

Following Krauss's suggestion in his article in Evans, *et al.* (1993) that 'we should refine and extend out analysis of two-level games by re-evaluating the roles of domestic actors' (Krauss, 1993, p. 293), and building on the work of Milner (1997) a clearer conceptualization of the domestic level than that used by Putnam (1988) in his 'Diplomacy and Domestic Politics' article was attempted to be put forward in Chapter 1. The conceptualization of the domestic level set out in Chapter 1 focuses on three variables – the state (bureaucracy), the legislature and other actors in the policymaking process – and the relationships between them. This conceptualization represents an alternative to using state-centered (Ikenberry *et al.*, 1988b, pp. 9–14), society-centered (Ikenberry *et al.*, 1988b, pp. 7–9), ideas-oriented (Goldstein, 1993), or rational choice (Milner, 1997, pp. 70–1) approaches to conceptualize the domestic level of trade policymaking when attempting to analyze the relationship between domestic politics and international relations in trade policymaking. When a state-centered conceptualization is used in an analysis of the relationship between domestic politics and international relations in trade policymaking, divisions between the state (bureaucracy), the legislature and other actors in the trade policymaking process are often under-emphasized. When a society-centered conceptualization is used, it is more difficult to analyze what is occurring within the state (bureaucracy) and among government officials. When an ideas-oriented conceptualization of the domestic level is used, it is harder to analyze evolving relations between domestic actors as Goldstein (1993, p. 11) admits, acknowledging that, without such analysis, 'policy analysis is incomplete'. When a rational choice approach is used, certain simplifying assumptions – such as that domestic actors are 'unitary and rational' or that the 'ideal policy is that preferred by the median voter' (Milner, 1997, pp. 33–4, 71) – make it difficult to analyze the potentially autonomous roles of the negotiator or legislator. Using such models, it is also difficult to analyze situations in which trade policy or the preferences of coalitions concerned with trade policy are different from those of the median voter.

Future research should question the universality of the neo-pluralist conceptualization of the domestic level of trade policymaking discussed in Chapter 1, and the application of this conceptualization to the analyses of the domestic politics of trade policymaking in various countries. While the all-inclusiveness of the 'other actors in the policymaking process' variable promotes parsimony by serving as a catch-all variable, this variable has the disadvantage of not distinguishing among certain clusters of interest groups, such as exporters and importers, consumers and producers, and the media. In future research, more specific hypotheses could be designed and tested, and particular case studies selected in order to clarify further the roles of such groups in trade policymaking.

At the domestic level, a central issue to reconsider is the concept of divided government and its relationship to the relative influence of the legislature in trade policymaking. Divisions can occur not only between the state (bureaucracy) and the legislature, but also within both the state (bureaucracy) and legislature. There may not always be a clear link between the degree to which a government is divided and the relative influence of the legislature, as the second hypothesis suggests. Chapters 3 and 4 demonstrate that the notion of 'division within a government' can be interpreted quite differently in a country, such as the United States, which is marked more by divisions between the Executive Branch and Congress, than in a county with a parliamentary government, such as Japan, where the Cabinet is generally composed of Diet members who are members of the political parties within the governing coalition. In future research, hypotheses could be designed and tested in order to examine the relative influence on trade policymaking of particular divisions between the state (bureaucracy) and the legislature, or particular divisions within the state (bureaucracy) or the legislature, following along the lines of the research in Martin (2000, pp. 48–52).

At the international level, more specific measurements, some of which could be borrowed from the literature on negotiations, could be developed to assess the extent to which cooperation is achieved in a particular trade negotiation. Such measurements might focus, following the work of Odell (2000, pp. 26–8), on the relationship between domestic politics and the development of the initial negotiating stances, the emergence of the resistance point between the actors involved, the emergence of a zone of agreement that comes to exist between the actors involved, the reaching of an agreement and its implementation. For example, during the Uruguay Round agriculture negotiations, domestic politics played an important role in the development of the initial negotiating stance of both the United States and Japan. Domestic politics also played an important role in the emergence of a resistance point – Japan's refusal to have the comprehensive tariffication of agricultural products applied to Japan's rice market – and in the development of the zone of agreement that emerged. The use of cooperation theory at the international level raises a question related to norms, concerning what constitutes cooperation in trade relations. According to Arthur Stein, in liberal versions of international trade theory, states are often treated 'as the primary units' and it is concluded 'that cooperative arrangements would emerge naturally from exchange' (Stein, 1990, p. 7). In such analyses, the influence of domestic politics is often not considered, and the extent to which a country cooperates is often simplified to an assessment of whether cooperation or defection occurs. Future studies along these lines might benefit from assessing the extent to which cooperation occurs by focusing more on the relationship between domestic politics and specific elements of the negotiating process. In addition, other hypotheses that focus on the international trade regime, the position of a country in the international political economy, and other international variables, could be developed and tested in order to explore further the relationship between domestic politics and international relations in trade policymaking.

As for analyzing the relationship between domestic politics and international relations in trade policymaking, Putnam's two-level game approach is specifically designed for particular types of negotiations – those that occur in two simplified stages: 'bargaining between the negotiators, leading to a tentative agreement', called Level I; and 'separate discussions within each group of constituents about whether to ratify the agreement', called Level II (Putnam, 1988, p. 436). Similarly, in Milner's model, the executive 'negotiate[s] with the foreign country and then the legislature ha[s] to ratify without amending the agreement' (Milner, 1997, p. 99).

In contrast, the contextual two-level game approach to analyzing trade policymaking does not separate the negotiation process so distinctly from the ratification process. The domestic level of the contextual two-level game approach, because it considers the relationship between the state (bureaucracy) and the legislature, allows for variations in the ratification processes of different countries, and for variations in ratification procedures for different types of negotiations. In the contextual two-level game approach, more emphasis is placed on the evolving relationship between domestic politics and negotiating strategies throughout a negotiation, and less emphasis is placed on the final act of ratification. This different focus permits more effective analyses of events such as the Uruguay Round agriculture negotiations,[1] where the legislatures and other actors in the policymaking process were in greater contact with the negotiators and more actively involved in the formation of the negotiating stances of their governments' representatives throughout the seven years of the negotiations than Putnam's two-level game approach suggests. However, it is important to note that Putnam's two-level game approach was largely developed in relation to the Bonn Summit conference of 1978, which was a very brief event in comparison to the Uruguay Round negotiations. The contextual two-level game approach to analyzing trade policymaking also allows for more effective analysis of the relationship between domestic politics and negotiating strategies during various stages of the negotiations, such as the preparation of preliminary drafts of a country's negotiating stance, or the preparation of draft agreements. During such stages, domestic political actors may be given a role in the negotiation, which is more complex than the single act of ratification at the end of a negotiation that Putnam's two-level game approach emphasizes.

The contextual two-level game approach also allows for the reality that the 'discussions within each group of constituents about whether to ratify the agreement' may not be 'separate' processes within each country, as Putnam's two-level game approach suggests. As Evans (1993, pp. 418–23) emphasizes, rather, the discussions within each country may be influenced during the course of the negotiation by transnational coalitions. Naka (1996, p. 126) states that the 'size (and strength) of transgovernmental coalitions' can provide 'a good reference for predicting the degrees of negotiations outcomes'.

The contextual two-level game approach to analyzing trade policymaking needs to be developed further through the analysis of the trade policymaking processes of countries other than the United States and Japan, and through the analysis of negotiations involving various economic sectors. Such research would make it possible to develop typologies of the relationship between domestic politics and international relations in various countries' trade policymaking processes, and to design and test other hypotheses concerning the relationship between domestic politics and international relations in trade policymaking.

Domestic politics and international relations in American trade policymaking

From Chapter 2 onwards, this book has attempted to develop a more systematic way of analyzing the relationship between domestic politics and international relations in American trade policymaking. Part of this approach involves conceptualizing the domestic level of American trade policymaking as consisting of the Executive Branch, Congress and other actors in the American trade policymaking process, and the relationships between them. This conceptualization of the domestic level of American trade policymaking represents an alternative to using state-centered (Ikenberry *et al.*, 1988b, pp. 9–14; Katzenstein, 1978b, pp. 308–13), society-centered (Pastor, 1980, pp. 61, 43–9), ideas-oriented (Goldstein, 1993), or rational choice (Robert Baldwin, 1976, pp. 7–15) approaches to conceptualize the domestic level in analyses of the relationship between domestic politics and international relations in American trade policymaking. Were a state-centered conceptualization used to analyze American trade policymaking during the Uruguay Round agriculture negotiations, it would be more difficult to analyze the important role that the Rice Millers' Association and other American agricultural interest groups played in pressuring the United States government to negotiate access to foreign agricultural markets. Were a society-centered conceptualization used, it would not be very easy to analyze evolving relations between the Executive Branch and the Congress, and within the Executive Branch during the negotiations. An ideas-oriented conceptualization would shift the focus of the analysis to how the ideas concerning global agricultural trade liberalization evolved during the negotiations, which would not necessarily lead to effective analysis of the evolving relationship between domestic politics and the American government's negotiating stance during the negotiations. Were a rational choice conceptualization of the domestic level used, the Executive Branch, Congress and other actors might be assumed to be unitary and rational. Direct causal links might be assumed to exist between voters' preferences and the evolution of the Executive Branch's negotiating stance, and between voters' preferences and the actions of members of Congress (Robert Baldwin, 1985, pp. 13–18). In some cases, it is difficult to find evidence of such direct links in the Uruguay Round agriculture negotiations. While it could be argued that members of Congress who represented districts producing particular agriculture products that could be affected by the Uruguay Round agriculture negotiations might have been motivated in some way by the fear that they might lose or gain votes by advocating a particular negotiating stance, the Executive Branch, rather than Congress, had ultimate control over the negotiating stances the United States government put forward in Geneva. While Congress could hold hearings, observe some negotiations, threaten not to extend fast-track negotiating authority, and express dismay over the Executive Branch's large degree of control over United States' negotiating stances, it did not exert much direct control over the content of the United States' negotiating stances, as Senator Baucus lamented in November 1989.[2]

Future research could develop and test hypotheses that focus on international factors other than cooperation, such as the international trade regime, the United States' position in the international political economy, and interdependence.

The contextual two-level game approach to analyzing American trade policymaking could also be developed further by examining case studies involving United States' trade relations with other countries, in other economic sectors and during other time periods.

Some of the strengths of the contextual two-level game approach to analyzing the relationship between domestic politics and international relations in American trade policymaking become evident if one compares the results of the analysis in Chapter 3 to previous efforts to use Putnam's two-level game approach to analyze the relationship between domestic politics and international relations in American trade policymaking during the Uruguay Round agriculture negotiations. Paarlberg (1993, 1997), for example, uses Putnam's two-level game approach to analyze American trade policymaking during the Uruguay Round agriculture negotiations focusing mostly on the negotiations concerning reducing domestic agricultural subsidies and agricultural export subsides. The negotiations concerning export subsidies were marked by constant pressure from agricultural interest groups and Congress, who sought to raise export subsidies in the event that an agreement was not reached. This was a quite different form of pressure from that created by agricultural exporters such as the Rice Millers' Association, who sought increased access to foreign markets, which resulted in the United States offering to place on the negotiating table some of its existing programs that restricted imports, a number of which were protected by the GATT waiver that the United States received in 1955. Paarlberg seeks to evaluate 'What was the payoff (or the penalty) for [the] decision to internationalize the domestic reform issue?' in the Uruguay Round negotiations, and concludes that 'the Uruguay Round was on balance a hindrance to unilateral agricultural policy reform' which 'added nothing to the pace or content of domestic farm policy reform in the United States (it may have actually delayed and weakened reform)' (Paarlberg, 1997, pp. 415, 434, 439). Paarlberg (1993, 1997) uses little of the terminology, such as coalitions, institutions or preferences, that Putnam suggests should be used to analyze the domestic level in two-level game analyses. Paarlberg describes a split that occurred at the beginning of the negotiations between the Reagan Administration's agricultural trade policymakers and agricultural interest groups dependent on subsidies. Paarlberg states that, partly because the Reagan Administration was frustrated by Congress's lack of desire to engage in 'deep, unilateral domestic farm support cuts', the Reagan Administration sought progress through internationalizing the problem by placing agriculture on the Uruguay Round negotiating agenda (Paarlberg, 1997, p. 419). Paarlberg states that the plan of US policymakers initially 'was to use an international negotiation, conducted with like-minded officials abroad, as a device both to weaken and then to finesse the domestic farm lobby opponents of reform at home'. Paarlberg argues that '[a]lmost from the moment that U.S. officials chose to describe agricultural reform as the make-or-break element in the larger GATT negotiations', US domestic agricultural interests 'began their own maneuvers to protect themselves from this internationalization' (Paarlberg, 1993, pp. 44–5).

In contrast to Paarlberg's analysis, Chapter 3 reveals that a primary concern of American policymakers when they placed agriculture on the Uruguay Round negotiating agenda was to counteract recent increases in the American trade deficit. Some policymakers in the Executive Branch and Congress were concerned not only with how to reduce American agricultural subsidies, but also with how to maintain and expand American agricultural exports by reaching an agreement with the EC that would lead to reductions in agricultural subsidies, and by reaching an agreement with Japan to improve access to agricultural markets. The analysis in Chapter 3 reveals a greater unity than Paarlberg's analysis suggests between the Executive Branch and America's large agricultural interest groups, such as the American Farm Bureau Federation, concerning the American government's negotiating stance. Chapter 3 reveals that, in the case of negotiations concerning agricultural market access, the primary driving force behind the US negotiating stance was a desire to increase American exports rather than to promote domestic reform. The domestic reform that occurred as a result of the agricultural market access negotiations – the reduction of certain American import restrictions – was agreed to largely in order to gain reciprocal access to other countries' agricultural markets, and was not developed, as Paarlberg suggests, primarily in response to the Reagan Administration's frustration with Congress's willingness to engage in 'deep, unilateral domestic farm support cuts' (Paarlberg, 1997, p. 419).

Domestic politics and international relations in Japanese trade policymaking

This book has also attempted to develop a more systematic way of analyzing the relationship between domestic politics and international relations in Japanese trade policymaking. Part of this approach involves conceptualizing the domestic level of Japanese trade policymaking by focusing on the bureaucracy, the Diet and other actors in the Japanese trade policymaking process, and the relationships between them. This conceptualization represents an alternative to using state-centered (Chalmers Johnson, 1982, pp. 17–34, 305–24), society-centered (Fukui, 1977, pp. 22–35), ideas-oriented (Pyle, 1996, p. 36), and rational choice (Ramseyer and Rosenbluth, 1993) conceptualizations of the domestic level in order to analyze the relationship between domestic politics and international relations in Japanese trade policymaking. Were a state-centered conceptualization to be used to examine the domestic level of Japanese trade policymaking, it would be difficult to analyze the interaction of various interest groups, such as the Central Union of Agricultural Co-operatives, with the bureaucracy and the Diet. Were a society-centered conceptualization used, it would be more difficult to analyze decisionmaking processes within the Japanese government, not only within the bureaucracy, but also within the Diet. As Chapter 4 demonstrates, Japan's various ministries, such as the Ministry of Agriculture, Forestry and Fisheries, the Ministry of International Trade and Industry, and the Ministry of Foreign Affairs, adopted positions concerning Japan's negotiating stance that resulted not just from pressure from various interest groups and Diet members, but also reflected to some extent perspectives developed within the Ministries themselves. Were an ideas-oriented approach used, the analysis of the domestic level would focus on how the idea of food security evolved and was promoted throughout the negotiations. While the idea of food security was a key concept in Japan's negotiating stance, analyzing the evolution of this idea during the negotiations would not necessarily lead to the effective analysis of the evolving relationship between domestic politics and Japan's nego-tiating stance during the Uruguay Round agriculture negotiations. Were a rational choice conceptualization used, the bureaucracy, Diet and other actors in the trade policymaking process might be assumed to be unitary and rational. Also, a direct link might be assumed to exist between voters' preferences and the evolution of Japan's negotiating stance. As revealed by various public opinion polls conducted during the negotiations, from the beginning of the Uruguay Round agricul-ture negotiations there was significant opposition to, as well as substantial public support for, altering Japan's virtual ban on rice imports. However, this ambivalence was not reflected in Japan's adam-ant opposition to the liberalization of Japan's rice market right up to the end of the negotiations.

Future studies of the relationship between domestic politics and international relations in Japanese trade policymaking could develop and test hypotheses that focus on international factors other than co-operation, such as the international trade regime, Japan's position in the international political economy, and interdependence. The ap-proach to analyzing Japanese trade policymaking elaborated in this book could also be developed further by carrying out additional case studies of Japanese trade relations with other countries in economic sectors other than agriculture during different time periods.

Some of the strengths of the approach to analyzing the relationship between domestic politics and international relations in Japanese trade policymaking developed in this book are revealed if one compares the results of the analysis in Chapter 4 to other efforts to use Putnam's two-level game approach to analyze the relationship between domestic pol-itics and international relations in Japanese trade policymaking during the Uruguay Round agriculture negotiations. David Rapkin and Aurelia George (1993) identify two domestic coalitions that emerged during the negotiations: first, a status quo coalition, composed of the Japanese gov-ernment, including bureaucrats from the Ministry of Agriculture, Forestry and Fisheries, the LDP, the opposition parties, the agricultural coopera-tives, most farmers, and consumer organizations; and second, a market-opening coalition composed of the Ministry of International Trade and

Industry, the Ministry of Foreign Affairs, and the Ministry of Finance, Japanese business federations, the media, consumers, and some farmers (Rapkin and Aurelia George, 1993, pp. 60–71). While Rapkin and George describe these two coalitions, they are less able, using Putnam's two-level game approach, to analyze the evolution of the coalitions during the negotiations. Rapkin and George's analysis places all of Japan's political parties in the status quo coalition, whereas Chapter 4 reveals that a progression of politicians and political parties, beginning with the political party Komeito during the summer of 1990, openly discussed various ways to liberalize Japan's rice market in order to conclude the Uruguay Round agriculture negotiations.

Rapkin and Aurelia George (1993) use Putnam's two-level game approach to 'analyze the intersecting domestic and international games of rice liberalization', emphasizing the 'factors affecting the range of potential bargains Japan would be able to strike' in the negotiations. Rapkin and George conclude that domestic constraints 'limited the Japanese government to reactive tactics and strategy, an abortive approach' that 'prevented initiatives that would allow self-definition of agricultural reform'. Rapkin and George also note that ratification procedures were 'somewhat ambiguous', particularly since the Uruguay Round had yet to be completed at the time they were writing (Rapkin and George, 1993, pp. 91–2).

In contrast, the analysis in Chapter 4 emphasizes the shifts that occurred within and among various Ministries, the Diet, and other actors in the Japanese trade policymaking process that affected the pro- and anti-liberalization coalitions and Japan's negotiating stance. Contrary to Rapkin and George's characterization of Japan's 'reactive tactics and strategy' (Rapkin and George, 1993, p. 91), Chapter 4 demonstrates that Japan undertook a relatively assertive, and not a reactive approach, from the beginning of the negotiations, as Japan's representatives argued for the creation of 'reliable and operationally effective rules that particularly provide fairer and clearer conditions for quantitative restrictions as exceptional measures' (Japan, Ministry of Agriculture, Forestry and Fisheries, 1987, pp. 13–14). Chapter 5 reveals that Japan was unable to find many countries that supported such calls for exceptional measures for its rice market. Chapter 4 reveals that the 'somewhat ambiguous' ratification process to which Rapkin and George refer, was in fact an ongoing process that occurred in various stages as Japan discussed the domestic implications of the April 1989 Mid-Term Review Agreement, de Zeeuw's July 1990 Framework Agreement, the December 1991 Draft Final Act, and ultimately the Final Act.

Yoichiro Sato also claims to use a two-level game approach to analyze Japan's role in the Uruguay Round agriculture negotiations. He argues that the negotiations leading up to Japan's December 1993 announcement of the partial liberalization of the rice market are an example of how 'complex domestic politics' 'may prevent state negotiators from taking a rational negotiating stance in the international arena'. Sato emphasizes that Japan's 'divided bureaucracy prevented flexible bargaining', and argues that 'Japan's adamant insistence on "food security" ' can be explained 'in terms of Japan's political inability to define "national interests" over sectoral cooperative interests of the bureaucracy and various private sectors' (Yoichiro Sato, 1996, pp. 96–7).

In comparison to Yoichiro Sato's work, Chapter 4 demonstrates the extent to which not only divisions within the bureaucracy but also divisions within the Diet and interest groups played a role in what Yoichiro Sato describes as Japan's inflexible bargaining position. Chapter 4 analyzes in greater detail the 'complex domestic politics' Sato describes, outlining the active debate that was occurring among bureaucrats, Diet members and interest groups concerning what limited exceptions to Japan's virtual ban on rice imports the country should offer in order to conclude the Uruguay Round agriculture negotiations. In contrast to Yoichiro Sato's description, Chapter 5 reveals that the negotiations concerning market access were part of a much larger negotiation, and that there was a certain rationale to Japan's seeking support from other countries for its negotiating stance. Holding out on consenting to increase access to Japan's rice market was used by Japan as a bargaining chip to force other countries to make concessions in other areas of the negotiations. Thus divisions not only within the bureaucracy but also within the Diet and among interest groups, as well as Japan's ability to delay making a concession in the negotiations, coupled with the length of time that the United States and the EC spent reaching an agreement, more than Japan's 'divided bureaucracy' alone, as Yoichiro Sato argues, may have been the reason for what he perceives as Japan's lack of flexible bargaining.

Domestic politics and international relations in US–Japan trade negotiations

By focusing on the relationship between domestic politics and cooperation and the interaction of negotiating strategies in US–Japan trade negotiations, this book has attempted to create a more rigorous analytic approach to the examination of the relationship between domestic politics and international relations in US–Japan trade negotiations, adding to the efforts of Krauss (1993), Schoppa (1997) and others to apply Putnam's two-level game approach to the analysis of US–Japan trade negotiations.

The approach to analyzing US–Japan trade negotiations developed in this book attempts to address Norio Naka's complaint that the state of the art in US–Japan trade negotiations studies 'essentially remain[s] compilations of case studies without systematically tested propositions based on clearly defined international relations theories'. Naka notes that many authors of studies of US–Japan trade negotiations 'have not been sufficiently self-critical of their conceptual models, the sets of assumptions that [were] included in them, nor the levels of analysis' (Naka, 1996, pp. xii, 225–6).[3] Few studies exist that set out clearly and debate the pros and cons of using state-centered, society-centered, ideas-oriented or rational choice conceptualizations in the analysis of the domestic level of US–Japan trade negotiations. (For one example of a study using a state-centered approach, see Katzenstein and Yutaka Tsujinaka (1995).)

While the hypotheses designed and tested in this book focus on co-operation and the interaction of negotiating strategies, future studies could examine, for example, the influence of the relationship between domestic politics and the international trade regime (as well as the international regime) on US–Japan trade negotiations (see Gilpin, 1988). In this regard, Schoppa notes the impact of the end of the Cold War and the conclusion of the Uruguay Round on US–Japan trade negotiations (Schoppa, 1999, pp. 318–23). The relationship between domestic politics and interdependence and its effect on US–Japan trade negotiations could also be examined. Hypotheses could be designed that examine the extent to which the relationship between domestic politics and stages of capitalism has affected US–Japan trade negotiations. Also, hypotheses that focus on the relationship between domestic politics and the impact of the use of various negotiating strategies, such as those outlined by Destler and Hideo Sato (1982b, pp. 279–87) and Schoppa (1997, pp. 32–45), could be developed and tested.

The approach to analyzing US–Japan trade negotiations described in this book needs to be developed further by examining other negotiations during other time periods and in other economic sectors, such as steel, automobiles and high technology. Such work could build on existing case studies – for example Curran (1982), Patrick and Hideo Sato (1982), Hideo Sato and Curran (1982), Hideo Sato and Michael Hodin (1982), and Winham and Ikuo Kabashima (1982).

Some of the strengths of the approach to analyzing the relationship between domestic politics and international relations in US–Japan trade negotiations developed in the book become evident if one compares the analysis in Chapter 5 to other efforts to use Putnam's two-level game approach to analyze the relationship between domestic politics and international relations in US–Japan trade negotiations during the Uruguay Round agriculture negotiations. For example, Rapkin and Aurelia George (1993, pp. 91–2), note the interaction of 'the radically liberal strategy pursued by the United States' with 'Japan's articulation of the food security doctrine', concluding that domestic constraints 'limited the Japanese government to reactive tactics and strategy, an abortive approach', which 'placed Japan in an isolated position'.

In contrast, this book develops and tests hypotheses concerning the extent to which domestic politics affected the agreement that the United States and Japan reached in the Uruguay Round agriculture negotiations, explaining in much greater detail the specific multilateral context at the GATT in Geneva in which the negotiations took place. Because of domestic politics, Japan adopted a protectionist negotiating stance in the Uruguay Round agricultural market acess negotiations, which made it difficult for Japan to cooperate with the United States and other countries in achieving an agreement concerning comprehensive tariffication of agricultural products and minimum access to agricultural markets. Because there was little division in the American government concerning the negotiating stance the United States took concerning comprehensive tariffication and minimum access, there was not much domestic pressure for the United States to alter its negotiating stance significantly during the negotiations. In contrast, greater division within the Japanese government concerning Japan's negotiating stance increased the influence of the Diet, which eventually made it easier for Japan to make some minor concessions in order to conclude the negotiations. Because several informed domestic groups in the United States and Japan endorsed the negotiating stances of the American and Japanese governments, the US Congress and the Japanese Diet were more likely to endorse the results of the negotiation, and the chances of the two countries reaching a cooperative agreement concerning comprehensive tariffication and minimum access increased.

In contrast to Yoichiro Sato's assertion that Japan's 'divided bureaucracy prevented flexible bargaining, allowing the U.S. negotiators to take advantage of the division' (Yoichiro Sato, 1996, p. 97), Chapters 4 and 5 demonstrate that Japan chose to adopt a protectionist negotiating stance early on in the agricultural market access negotiations, and to maintain that stance while a lively domestic debate occurred, until a last-minute compromise was reached which required the United States and Japan to make concessions in their negotiating stances to conclude the Uruguay Round negotiations.

* * *

The contextual two-level game approach to analyzing trade policymaking described in this book represents an attempt not only to improve on Putnam's two-level game approach, but also to fill a gap in the literature concerning the relationship between domestic politics and international relations in trade policymaking. While the research presented here represents an essential step towards increasing the understanding of American trade policymaking, Japanese trade policymaking and US–Japan trade negotiations, the contextual two-level game approach to analyzing trade policymaking needs to be developed further by applying it to other countries and other negotiations. An effective starting point for such future study would be the analysis of a series of case studies that illuminate the relationship between domestic politics and international relations in trade policymaking in several major countries during the postwar period.

Notes

Preface and Acknowledgements

1   Lists of derestricted and still restricted Uruguay Round documents are contained in Annex I and Annex II of GATT, Uruguay Round Trade Negotiations Committee, 'Derestriction of Uruguay Round Documents: Documents Proposed for Derestriction on 3 January 1997', GATT Doc. No. MTN.TNC/46, 31 October 1996. The notice of the derestriction of GATT Doc. No. MTN. TNC/ 46 and the documents listed in Annex I of GATT Doc. No. MTN.TNC/46 is GATT, Uruguay Round Trade Negotiations Committee, 'Derestriction of Uruguay Round Documents: Documents derestricted on 3 January 1997', GATT Doc. No. MTN.TNC/47, 6 January 1997. See also GATT, Uruguay Round Trade Negotiations Committee, 'Derestriction of Uruguay Round Documents: Documents derestricted on 3 January 1997: Addendum', GATT Doc. No. MTN. TNC/47/Add.I, 16 September 1997.

1 Introduction

1   Caporaso (1997) identifies three different ways of exploring the relationship between domestic politics and international relations: two-level games, as exemplified by Putnam, 1988; the second image reversed, as exemplified by Ronald Rogowski, 1989; and Caporaso's (1997) domestification of international politics approach.
2   See Ikenberry, 1988, pp. 219, 222–9; see also Ikenberry et al., 1988b, pp. 9–14 and Friman, 1990, pp. 15–19. And see, for example, Matthew Evangelista, 1997; Katzenstein, 1978b; Krasner, 1976; Krasner, 1978b; Thomas Risse-Kappen, 1995.
3   See, in general, Milner, 1997, p. 71. See also Bruce Bueno de Mesquita, 2000, pp. 384–7; Milner, 1997, pp. 99–128.
4   (Bureaucracy) is inserted in parantheses after 'the state' in order to separate this conceptualization from the notion of the state used in Evans et al., 1985.
5   See, similarly, Milner's use of polyarchy in the domestic level of the model developed in Milner (1997, pp. 11–12), and the discussion of the evolving relationship between socioeconomic sectors, institutions, and public policy paradigms in Pempel (1998, pp. 22–7). Unlike Milner's conceptualization of the domestic level, which includes political actors (executive and legislature) and societal actors (interest groups), which are assumed to be 'unitary and rational' (see Milner, 1997, pp. 33–4), in the contextual two-level game approach to analyzing trade policymaking, the state (bureaucracy), the legislature and other actors in the policymaking are not considered to be unitary and rational.

6 For a different approach, see the categorization of countries as being either democracies composed of an executive and a legislature, or autocracies composed of a unitary actor in Edward Mansfield *et al.*, 2000, p. 306.

7 See Ikenberry *et al.*, 1988b, pp. 4–5. See, in general, Gilpin, 1987, pp. 85–92; see also Peter Cowhey and Edward Long, 1983; Keohane, 1984; Snidal, 1985.

8 See, in general, Gourevitch, 1978b, pp. 892–3. Gourevitch cites as examples Keohane and Joseph Nye, 1971; Edward Morse, 1976. See also Keohane and Nye, 1989.

9 See, in general, Gourevitch, 1978b, pp. 888–91. Gourevitch cites as examples Fernando Henrique Cardoso, 1973; Andre Gunder Frank, 1967; Gourevitch, 1978a; Wallerstein, 1974; see also Evans, 1979.

10 See, in general, Gourevitch, 1978b, p. 885–8. Gourevitch cites as examples David Collier, 1979; Alexander Gerschenkron, 1962; Moore, 1966; Guillermo O'Donnell, 1974.

11 See, in general, Kenneth Oye, 1985b, pp. 18–22; Snidal, 1991, pp. 716–19. See also Milner, 1992, pp. 473–4; Schoppa, 1997, pp. 40–2.

12 See I. William Zartman, 1994b. See also John Conybeare, 1987, pp. 21–72; Beth Yarbrough and Robert Yarbrough, 1992.

13 See, in general, Hampson, 1995, pp. 3–51; Victor Kremenyuk, 1991; James March, 1994; Odell, 2000, pp. 24–46; Howard Raiffa, 1982; Zartman, 1994a.

14 For one example of a typology of different countries' trade policymaking processes, see Pat Choate and Juyne Linger, 1988. See also the discussion of liberalism and mercantilism in Gilpin, 1987, pp. 172–83; Bruce Moon, 1996, pp. 23–52; see also Mansfield *et al.*, 2000, p. 306.

15 These three hypotheses are derived from Milner, 1997, pp. 242, 98 and 239–40.

2 Domestic Politics and International Relations in US–Japan Trade Policymaking

1 See in general, Ikenberry *et al.*, 1988b, pp. 9–14; Katzenstein, 1978b, pp. 308–13; Krasner, 1978a; see also the institution-based approach in Sharyn O'Halloran, 1994.

2 See Stefanie Lenway, 1985, pp. 21–42; Pastor, 1980, pp. 43–9, 61; see also Aaronson, 1996; Jeff Frieden, 1988.

3 This neo-pluralist vision of Japanese trade policymaking is based on Muramatsu and Krauss's patterned pluralist model of Japanese policymaking discussed in Muramatsu and Krauss, 1987, pp. 537–9, 542–3; as well as in Muramatsu, 1993. For more general discussions of the pluralist model of Japanese policymaking, see Gary Allinson, 1989; Fukui, 1977, pp. 35–48; Schwartz, 1998, pp. 40–7.

4 More in-depth research into the functioning of such coalitions should make reference to research concerning the role of networks. See, for example, Jeffrey Broadbent and Yoshito Ishio, 1998.

5 See Okimoto, 1988, pp. 318–23; see also Destler *et al.*, 1976, pp. 68–86; Chikara Higashi, 1983, pp. 36–46.

6 'The Constitution of Japan' (*Nihonkoku Kempō*) in Kyoko Inoue, 1991, pp. 290–1.
7 Concerning the evolution of the GATT negotiating rounds, see Gilpin, 1987, pp. 190–203; Jackson, 1989, pp. 52–7.
8 By focusing on the United States and Japan, however, this book is in no way advocating a US–Japan relationship to the exclusion of the interests of other countries, such as developing countries.

3 The US and the Uruguay Round Agriculture Negotiations

1 Agricultural Adjustment Act, amendments, US Statutes at Large 49, 1936, p. 774; see also 'Limitations on imports; authority of President', *US Code*, vol. 7, sec. 624, 1988.
2 Jimmye Hillman, 1991, p. 67. See also C. Fred Bergsten *et al.*, 1987, pp. 206–17.
3 Hillman, 1991, p. 67. Concerning import restrictions on sugar, dairy products and peanuts, see Gary Hufbauer *et al.*, 1986, pp. 286–322.
4 'Testimony of Ambassador William E. Brock, III, U.S. Trade Representative', in US Congress, Senate Committee on Finance, Subcommittee on International Trade, 1 March 1982, pp. 25–33, 33.
5 'Statement of Lionel Olmer, Under Secretary for International Trade, Department of Commerce', in US Congress, Senate Committee on Banking, Housing, and Urban Affairs, Subcommittee on International Finance and Monetary Policy, 4 March 1982, pp. 4–9, 5. For further details, see 'Statement of David Macdonald, Deputy U.S. Trade Representative', in US Congress, Senate Committee on Banking, Housing, and Urban Affairs, Subcommittee on International Finance and Monetary Policy, 4 March 1982, pp. 10–14.
6 'Prepared Statement of Hon. William E. Brock', in US Congress, Joint Economic Committee, 3 February 1983, pp. 6–29, 25.
7 'Debate over Reciprocity, U.S. Market Access Subject of Tennessee Conference', *International Trade Reporter's U.S. Import Weekly*, 23 February 1982, p. 570.
8 'Response of William E. Brock, U.S. Trade Representative', in US Congress, House Committee on Appropriations, Subcommittee on the Departments of Commerce, Justice, and State, the Judiciary and Related Agencies, 27 February 1985, p. 293.
9 US Congress, House Committee on Agriculture, 1985, pp. 8–9.
10 Ibid., pp. 70–1.
11 'Statement of Allan I. Mendelowitz, Senior Associate Director, National Security and International Affairs Division, U.S. General Accounting Office', in US Congress, House Committee on Agriculture, Subcommittee on Department Operations, Research, and Foreign Agriculture, 29 September 1986, pp. 31–5, 33.
12 US Congress, House Committee on Agriculture, 1985, pp. 44–5.

13 'Response of Daniel Amstutz, Agriculture Under Secretary', in US Congress, House Committee on Foreign Affairs Subcommittee on Europe and the Middle East, 24 July 1986, p. 81.
14 'Remarks at a White House Meeting with Business and Trade Leaders, September 23, 1985', in Reagan, *1985, Book II – June 29 to December 31, 1985*, 1988, p. 1129.
15 'Statement of Daniel G. Amstutz, Under Secretary, International Affairs and Commodity Programs, U.S. Department of Agriculture', in US Congress, House Committee on Agriculture, 15 April 1986, pp. 5–7, 5–6.
16 'Statement of Francis B. Gwin, Member, National Commission on Agricultural Trade and Export Policy', in US Congress, Senate Committee on Agriculture, Nutrition, and Forestry, Subcommittee on Foreign Agricultural Policy, 17 June 1986, pp. 331–42, 333–5.
17 Ambassador Clayton Yeutter, United States Trade Representative, 'Speech to the United States Chamber of Commerce, September 10, 1986', US Trade Representative, Speeches and Testimonies File, 1986, p. 4.
18 'Punta del Este Declaration, Ministerial Declaration of 20 September 1986', in GATT, *GATT Activities 1986*, 1987, p. 21.
19 'U.S. Rice Group Files Complaint against Japan, Argentina urges USTR to Drop Soybean Case', *International Trade Reporter*, 17 September 1986, p. 1126.
20 'Prepared Statement of C. Ronald Caffey', in US Congress, Senate Committee on Finance, Subcommittee on International Trade, 3 November 1989, pp. 34–6, 34.
21 'USTR Rejects Complaint Against Japan's Rice Import Curbs, Will Pursue Issue in New Round', *International Trade Reporter*, 29 October 1986, p. 1304.
22 'Fifth Meeting of the Group of Negotiations on Goods: Record of Decisions Taken', GATT, Uruguay Round Group of Negotiations on Goods, GATT Doc. No. MTN.GNG/5, 9 February 1987, pp. 3–4, 10–11.
23 See 'List of Representatives', GATT, Uruguay Round Negotiating Group on Agriculture, GATT Doc. No. MTN.GNG/NG5/INF/1, 26 October 1987, pp. 18–19.
24 See 'Derestriction of Uruguay Round Documents: Documents proposed for derestriction on 3 January 1997', GATT, Uruguay Round Trade Negotiations Committee, GATT Doc. No. MTN.TNC/46, 31 October 1996; 'Derestriction of Uruguay Round Documents: Documents derestricted on 3 January 1997', GATT, Uruguay Round Trade Negotiations Committee, GATT Doc. No. MTN.TNC/47, 6 January 1997.
25 'Testimony of Suzanne Early, Assistant U.S. Trade Representative', in US Congress, House Committee on Agriculture, Subcommittee on Livestock, Dairy, and Poultry, 4 March 1987, pp. 211–14, 213.
26 US Congress, House Committee on Agriculture, 7 April 1987, p. 119.
27 Ibid., pp. 64–5.
28 US Congress, Congressional Budget Office, 1987, pp. 74, 77, 93.
29 'United States Proposal of 6 July 1987 for Negotiations on Agriculture', in Ernst-Ulrich Petersmann and Meinhard Hilf, 1988, p. 585.
30 'Major Farming Nations Applaud U.S. Subsidy Reform Plan, Yet Question its Feasibility', *International Trade Reporter*, 8 July 1987, p. 861.
31 Omnibus Trade and Competitiveness Act of 1988, US Statutes at Large 102, 1988, pp. 1395–6; US Congress, House Conference Rept, 1988, pp. 874–7.

32 'U.S. Rice Industry Files Section 301 Request Against Japanese Import Barriers to U.S. Rice', *International Trade Reporter*, 21 September 1988, p. 1268.
33 'Yeutter Rejects Rice Industry 301 Petition, Says GATT is a Better Forum for Complaint', *International Trade Reporter*, 2 November 1988, pp. 1442–3, 1442.
34 'Trade in Agricultural Products . . . 11–14 October', GATT, *News of the Uruguay Round of Multilateral Trade Negotiations*, NUR 020, 4 November 1988, p. 7.
35 'Remarks to the National Chamber Foundation, November 17, 1988', in Reagan, *1988–89, Book II – July 2, 1988 to January 19, 1989*, 1991, pp. 1522–3.
36 'Dunkel Plan Would Prevent Use of AD/CVD Laws Against Unfair Traders, NFU Charges', *International Trade Reporter*, 5 April 1989, p. 417.
37 'List of Representatives', GATT Uruguay Round Trade Negotiations Committee, GATT Doc. No. MTN.TNC/INF/6/Rev.1, 6 April 1989, pp. 59–63.
38 'Trade Negotiations Committee Meeting at Level of High Officials, Geneva, 5–8 April 1989', GATT Uruguay Round Trade Negotiations Committee, GATT Doc. No. MTN.TNC/9, 11 April 1989, pp. 3–6.
39 'Statement of Irvin J. Elkin, President, Associated Milk Producers, Inc., Amery, WI', in US Congress, Senate Committee on Finance, 20 April 1989, pp. 11–12, 11.
40 'Responses by Ann Veneman, Associate Administrator, Foreign Agricultural Service, and Suzanne Early, Assistant U.S. Trade Representative, to Questions Submitted by Senator Kent Conrad', in US Congress, Senate Committee on Agriculture, Nutrition, and Forestry, Subcommittee on Domestic and Foreign Marketing and Product Promotion, 12 April 1989, pp. 23–5, 24.
41 'Agriculture . . . 10 to 12 July', GATT, *News of the Uruguay Round of Multilateral Trade Negotiations*, NUR 030, 3 August 1989, p. 8.
42 'Growing Awareness of Farm Subsidy Costs Resulted in Tariffication Plan, Hills Says', *International Trade Reporter*, 19 July 1989, p. 951.
43 'Agriculture . . . 25–26 October', GATT, *News of the Uruguay Round of Multilateral Trade Negotiations*, NUR 032, 21 November 1989, p. 6.
44 'Opening Statement of Hon. Max Baucus, a U.S. Senator from Montana, Chairman of the Subcommittee', in US Congress, Senate Committee on Finance, Subcommittee on International Trade, 3 November 1989, pp. 1–4, 2–3.
45 'Prepared Statement of Senator Max Baucus', in US Congress, Senate Committee on Finance, Subcommittee on International Trade, 3 November 1989, pp. 33–4, 33.
46 'Response of Hon. Julius L. Katz, Ambassador, Deputy U.S. Trade Representative', in US Congress, Senate Committee on Finance, Subcommittee on International Trade, 3 November 1989, p. 9.
47 'Statement of C. Ronald Caffey, Director, Raw Materials, Uncle Ben's, Inc., Testifying on Behalf of the Rice Millers' Association, Houston, TX', in US Congress, Senate Committee on Finance, Subcommittee on International Trade, 3 November 1989, pp. 22–4, 23.
48 'Prepared Statement of Eiler C. Ravnholt, Vice President and Washington Representative, Hawaiian Sugar Planters Association, testifying on behalf of the U.S. Sugar Cane Growers and Processors, and the U.S. Sugar Beet Growers and Processors', in US Congress, Senate Committee on Finance, 3 November 1989, pp. 44–8, 45.

49 'Statement of C. Ronald Caffey, Chairman of the Board, Rice Millers' Association, and Director of Raw Materials, Uncle Ben's, Inc., Houston, TX', in US Congress, Senate Committee on Agriculture, Nutrition, and Forestry, Subcommittee on Agricultural Production and Stabilization of Prices, 5 March 1990, pp. 163–5, 163.

50 US International Trade Commission, 1990b, pp. 1–5, 135–41.

51 'Framework Agreement on Agriculture Reform Programme, Draft Text by the Chairman', GATT Uruguay Round Negotiating Group on Agriculture, GATT Doc. No. MTN.GNG/NG5/W/170, 11 July 1990, pp. 1, 2, 4.

52 'Statement of Suzanne Early, Assistant U.S. Trade Representative for Agriculture', in US Congress, House Committee on Agriculture, Subcommittee on Department Operations, Research, and Foreign Agriculture, 28 February 1991, pp. 5–7, 6.

53 President, Proclamation, 'Modification of Tariffs and Quota on Certain Sugars, Syrups, and Molasses, Proclamation 6179', *Federal Register*, 55, 13 September 1990, p. 38,293.

54 'Bush Converts U.S. Sugar Quota Program to GATT-legal Tariff-Rate Quota Scheme', *International Trade Reporter*, 19 September 1990, p. 1431.

55 'Sugar Alliance Says It Can No Longer Support "Weakened" U.S. Position in GATT Negotiations', *International Trade Reporter*, 24 October 1990, p. 1620.

56 'U.S. Advances Final Farm Proposal Calling for 90 Percent Cuts in Export Subsidy Levels', *International Trade Reporter*, 17 October 1990, pp. 1564–5. See also 'U.S. Proposal for Agricultural Negotiations in the Uruguay Round of Multilateral Trade Talks with Accompanying Fact Sheet Released by the Office of the U.S. Trade Representative Oct 15, 1990', *International Trade Reporter*, 17 October 1990, pp. 1594–9.

57 *Omnibus Budget Reconciliation Act of 1990*, US Statutes at Large, 104, 1990, pp. 1388–12 to 1388–14.

58 'Negotiations to Liberalize World Trade Stall as EC Stands Firm on Farm Offer', *International Trade Reporter*, 5 December 1990, pp. 1820–1, 1821.

59 'Uruguay Round of Multilateral Trade Talks Tops U.S. Trade Agenda for 1991 as Administration, Congress also Prepare to Deal with Range of Other Issues', *International Trade Reporter*, 9 January 1991, p. 60.

60 'Response of Suzanne Early, Assistant U.S. Trade Representative for Agriculture', in US Congress, House Committee on Agriculture, Subcommittee on Department Operations, Research, and Foreign Agriculture, 28 February 1991, pp. 7–8.

61 'Statement of Allan I. Mendelowitz, Director, International Trade, Energy, and Finance Issues, National Security and International Affairs Division, U.S. General Accounting Office', in US Congress, House Committee on Agriculture, 28 February 1991, pp. 52–4, 54.

62 'Response of Ambassador Hills', in US Congress, House Committee on Agriculture, 13 March 1991, p. 22.

63 'Statement of the American Farm Bureau Federation', in US Congress, House Committee on Agriculture, 13 March 1991, pp. 121–9, 121, 125, 127.

64 'Statement of Rice Millers' Association', in US Congress, House Committee on Agriculture, 13 March 1991, pp. 209–10, 210.

65 'Statement of David Senter, National Director, American Agriculture Movement, Inc.', in US Congress, House Committee on Agriculture, 13 March 1991, pp. 211–18, 213–14.

66 'Statement of Stewart G. Huber, President, Farmers Union Milk Marketing Cooperative', in US Congress, House Committee on Agriculture, 13 March 1991, pp. 205–8, p. 206.

67 'Appendix B, NFU 1991 Policy Statement on International Trade Pacts', in US Congress, House Committee on Agriculture, 13 March 1991, pp. 183–5, 185.

68 'Response of Ambassador Hills', in US Congress, Senate Committee on Agriculture, Nutrition, and Forestry, 8 May 1991, p. 50.

69 'Response to Questions from Senator [Larry] Craig', in US Congress, Senate Committee on Agriculture, Nutrition, and Forestry, 8 May 1991, pp. 124–32, 130–1.

70 'Senate and House Vote to Extend Fast Track for North American FTA, Uruguay Round Talks', *International Trade Reporter*, 29 May 1991, pp. 802–4, 802.

71 'Statement of Julius L. Katz, Deputy United States Trade Representative, Office of the United States Trade Representative', in US Congress, House Committee on Agriculture, 10 December 1991, pp. 13–16, 14–15.

72 'Statement of Randolph Nodland, Past President, on Behalf of the National Family Farm Coalition', in US Congress, House Committee on Agriculture, 10 December 1991, pp. 68–71, 70–1.

73 'Statement of Harvey Schneider, President, Southeast Peanut Growers Association, on Behalf of James Earl Mobley, Chairman, National Peanut Growers Group, accompanied by Billy Bain', in US Congress, House Committee on Agriculture, 10 December 1991, pp. 54–6, 54.

74 'Paragraph 4 and Annex 3, Market Access, Section A: The Calculation of Tariff Equivalents and Related Provisions, Paragraph 1 of Part B: Agreement on Modalities for the Establishment of Specific Binding Commitments under the Reform Programme, L. Text on Agriculture', in 'Draft Final Act Embodying the Results of the Uruguay Round of Multilateral Trade Negotiations, GATT, Uruguay Round Trade Negotiations Committee, GATT Doc. No. MTN.TNC/W/FA, 20 December 1991, pp. L. 19, 25.

75 'Paragraph 5, Part B, Agreement on Modalities for the Establishment of Specific Binding Commitments under the Reform Programme, L. Text on Agriculture', in 'Draft Final Act', GATT, Uruguay Round Trade Negotiations Committee, GATT Doc. No. MTN.TNC/W/FA, 20 December 1991, p. L. 19.

76 'Response of Richard Crowder, Agriculture Deputy Under Secretary for International Affairs and Commodity Programs', in US Congress, House Committee on Agriculture, 9 January 1992, pp. 377, 476–7.

77 'Statement of Dean R. Kleckner, President, American Farm Bureau Federation', in US Congress, House Committee on Agriculture, 9 January 1992, pp. 440–3, 442.

78 'Statement of Julius L. Katz, Deputy United States Trade Representative, Office of the United States Trade Representative', in US Congress, House Committee on Agriculture, 31 March 1992, pp. 697–8, 697.

79 'President Bush, Other G-7 Leaders Expect GATT Trade Agreement by End of this Year', *International Trade Reporter*, 8 July 1992, p. 1193; 'U.S. Rice Industry Prepares Section 301 Petition Against Japan', *International Trade Reporter*, 18 November 1992, p. 1961.

80 GATT, *GATT Activities 1992*, 1993, p. 18.
81 'Responses of Mr. Kantor to Questions Submitted by Senator [Tom] Daschle', in US Congress, Senate Committee on Finance, 19 January 1993, pp. 60–1.
82 'U.S. Will Push for End to Rice Bans in Japan and Korea, USDA's Espy Says', *International Trade Reporter*, 3 March 1993, p. 365.
83 Ibid.
84 'Testimony of Ambassador Mickey Kantor, United States Trade Representative', in US Congress, House Committee on Agriculture, 17 March 1993, pp. 85–8, 87–8.
85 'Response of Ambassador Kantor', in US Congress, House Committee on Agriculture, 17 March 1993, p. 60.
86 'Statement of Hon. Mickey Kantor, U.S. Trade Representative', in US Congress, Senate Committee on Finance, 20 May 1993, pp. 7–10, 8.
87 'Response of Ambassador Kantor', in US Congress, Senate Committee on Finance, 20 May 1993, p. 22.
88 Extension of Uruguay Round Trade Agreement Negotiating and Proclamation Authority and of 'Fast Track' Procedures to Implementing Legislation, US Statutes at Large, 107, 1993, pp. 239–40.
89 GATT, *GATT Activities 1993*, 1994, p. 16.
90 'Report on the Uruguay Round', GATT, Uruguay Round Trade Negotiations Committee, GATT Doc. No. MTN.TNC/W/113, 13 July 1993, p. 2.
91 'Statement of Hon. Michael Kantor, U.S. Trade Representative', in US Congress, House Committee on Ways and Means, Subcommittee on Trade, 13 July 1993, pp. 6–8, p. 6.
92 'Report from the G-7 Economic Summit, Testimony to the House Subcommittee on Trade Ways and Means Committee, Ambassador Michael Kantor, U.S. Trade Representative', in US Congress, House Committee on Ways and Means, Subcommittee on Trade, 13 July 1993, pp. 9–12, 10.
93 'Statement and Testimony of David L. Senter, Director of Congressional Affairs, American Corn Growers Association, also on Behalf of National Farmers Union, and National Family Farm Coalition', in US Congress, House Committee on Ways and Means, Subcommittee on Trade, 4–5 November 1993, pp. 231–41, 231, 234.
94 'Progress Made in U.S.–EC Trade Negotiations as of Dec. 7 Agriculture', *International Trade Reporter*, 8 December 1993, p. 2042.
95 See comparison in Breen, 1999, pp. 61–165.
96 'The Final Act of the Uruguay Round: Press Summary', GATT, *News of the Uruguay Round of Multilateral Trade Negotiations*, NUR 080, 14 December 1993, pp. 8–9.
97 'Uruguay Round Agreement is Reached', *International Trade Reporter*, 15 December 1993, p. 2103.
98 'Letter to Congressional Leaders on the General Agreement on Tariffs and Trade', in Clinton, *1993, Book II – August 1 to December 31, 1993*, 1994, pp. 2180–2, 2180.
99 See 'Uruguay Round Agreement is Reached', *International Trade Reporter*, 15 December 1993, p. 2103.
100 See GATT, 1994b.

101 'Response of Secretary Espy', in US Congress, House Committee on Agriculture, 16 March 1994, p. 87.
102 'Statement of Robert Watts, Vice President, Commodity and International, Riviana Foods, on Behalf of the Rice Millers' Association and U.S. Rice Producers' Group', in US Congress, House Committee on Agriculture, 16 March 1994, pp. 169–70.
103 'Statement of David Senter, Director, Congressional Affairs, American Corn Growers Association', in US Congress, House Committee on Agriculture, 20 April 1994, pp. 172–4, 173.
104 See US Congress, House, Message from the President of the United States Transmitting the Uruguay Round Trade Agreements, Texts of Agreements, Implementing Bill, Statement of Administrative Action and Required Supporting Statements, 1994.
105 Statement of Administrative Action of the Uruguay Round Agreements Act concerning the 'Agreement on Agriculture', in US Congress, House, Message from the President of the United States Transmitting the Uruguay Round Trade Agreements, 1994, p. 712.
106 Ibid., pp. 712–13.
107 'Statement of Hon. Mike Espy, Secretary of Agriculture, United States Department of Agriculture', in US Congress, Senate Committee on Agriculture, Nutrition, and Forestry, 20 April 1994, pp. 18–23, 19.
108 For one analysis of these provisions, see Breen, 1993, pp. 45–8.
109 See US Congress, Senate Committee on Finance, Committee on Agriculture, Nutrition, and Forestry, Committee on Governmental Affairs, 22 November 1994, pp. 126–31, 203–20.

4   Japan and the Uruguay Round Agriculture Negotiations

1 'Taibei Nōsanbutsu Kōshō mo Sakiokuri? "Jōho Hantai" no Yatō Fujō de' (Negotiations concerning Agricultural Products with the United States, What Lies Ahead Next? Opposition Parties 'Oppose Concessions' on the Surface), *Mainichi Shimbun*, 20 December 1983, p. 9.
2 See 'Shūin mo Kome Jikyū Ketsugi' (Lower House As Well Adopts Rice Self-Sufficiency Resolution), *Asahi Shimbun*, 21 July 1984, p. 9; 'Both Houses Adopt Resolution for Rice Self-Sufficiency', *Japan Agrinfo Newsletter*, vol. 2, no. 1, August 1984, p. 3.
3 'Kanzei Bubun no Kōdō Keikaku Naiyō' (Tariff Part of Action Program Contents), *Nihon Keizai Shimbun*, 25 June 1985, p. 2.
4 'The First "Maekawa" Report, The Report of the Advisory Group on Economic Structural Adjustment for International Harmony, Submitted to the Prime Minister April 7, 1986', in Choy, 1988, pp. 9, 11. See also 'Keikōken Hōkoku, Kuroji Taishitsu Kaizen e Seisaku Tenkan' (Economic Structure Research Report, Policy Changes to Improve Surplus Situation), *Nihon Keizai Shimbun*, 8 April 1986, p. 1.
5 'Jimin Shūin, Tsuika Kōnin Fukume 304 Giseki' (LDP Lower House, With Additional LDP Nominations Includes 304 Diet Seats), *Nihon Keizai Shimbun*, 8 July 1986, p. 1.

6 See 'Seisansha Beika, Sueoki Kettyaku, Jimin Teikō, Hikisage Hanekaesu, Shushō ga Saishū Ketsudan, Ōhabana Seiji Kasan 3.8%' (Rice Producers' Price, Decision to Leave As Is, LDP Resistance, Reduction Driven Back, Prime Minister Makes Final Decision, Large Political Addition of 3.8%), *Nihon Keizai Shimbun* (evening edition), 9 August 1986, p. 1.

7 'Beikoku Seifu no Dekata Chūshi, Kome Teiso de Nōsuishō · Dantai' (Closely Observing the American Government's Attitude Concerning the Rice Petition, the Ministry of Agriculture, Forestry and Fisheries and Interest Groups), *Asahi Shimbun*, 10 September 1986, p. 9.

8 'Honsha Yoron Chōsa, "Shokkansei Haishi" wa 18%, Kome Jiyūka Sanpi Mapputatsu' (*Mainichi Shimbun* Public Opinion Poll, 'Abolish the Food Control System' 18%, Concerning Rice Liberalization, Divided down the Middle), *Mainichi Shimbun*, 16 December 1986, p. 3.

9 'Prime Minister Shows Strong Intentions on Farm Reform', *Japan Agrinfo Newsletter*, vol. 4, no. 5, December 1986, p. 4.

10 'The Second "Maekawa" Report: Summary Report of the Economic Council's Special Committee on Economic Restructuring, April 23, 1987', in Choy, pp. 16–18. See also 'Sannen Inai ni Seisaku Doryoku Shūchū, Shokkansei Minaoshi, Jūtaku Kyōkyū Sokushin, Rōdō Jikan Tanshuku, Shin "Maekawa Ripōto" Saishū Hōkoku no Zenyō' (Policy Efforts are Concentrated Within Three Years, Reexamine Food Control System, Promote Housing Supply, Shorten the Work Week, the Entire Contents of the Last Version of the New 'Maekawa Report'), *Nihon Keizai Shimbun*, 22 April 1987, p. 1.

11 See 'List of Representatives', GATT, Uruguay Round Negotiating Group on Agriculture, GATT Doc. No. MTN.GNG/NG5/INF/1, 26 October 1987, pp. 11–12.

12 See 'Shushō Shoshin Hyōmei Enzetsu no Naiyō' (Content of Prime Minister's Speech Expressing his Beliefs), *Nihon Keizai Shimbun*, 7 July 1987, p. 5; as summarized and translated in 'Policy Speech by Prime Minister Yasuhiro Nakasone', *Japan Agrinfo Newsletter*, vol. 5, no. 1, August 1987, p. 2.

13 'Text of Economic Declaration by 7 Leading Industrial Nations', *New York Times*, 11 June 1987, sec. A, p. 16.

14 'Shin Raundo, Nōgyō Bōeki no Gensoku Teian, Nihon "Tokushusei" Uttae' (Uruguay Round, Proposal concerning Principles of Agricultural Trade, Japan Appeals concerning 'Special Characteristics'), *Nihon Keizai Shimbun* (evening edition), 7 July 1987, p. 1.

15 'Agriculture', in Japan Institute of International Affairs, 1989, p. 125.

16 L/6253, adopted on 2 February 1988, 35S/163, 229, paras. 5.2.2.1–2 (GATT, 1994c, p. 295).

17 'Nōsanbutsu Jiyūka ni Hantai, Zenchū · Jimin Nōrin Giin ga Ketsugi' (Opposing the Liberalization of Agricultural Products, the Central Union of Agricultural Co-operatives and the LDP Agriculture and Forestry Diet Members Resolve), *Nihon Keizai Shimbun*, 6 November 1987, p. 5.

18 'Gatto Saiteian, Ōsuji Ukeire e, Seifu Hōshin, Taibei Kōshō Fuchō nara; Nyū Seihin ya Denpun, Arata na Hogosaku Kentō' (GATT Decision Plan, Towards Accepting Almost Everything, Government Policy, if Negotiations with the United States End Badly; Dairy Products and Starch, New Protectionist Policy Considered), *Asahi Shimbun*, 2 November 1987, p. 3.

19 'Nōgyō Shin Raundo Nihon Teian, Reigaiteki ni Yunyū Seigen mo, EC no Yushutsu Hojo Teppai Yōkyū' (Japan's Proposal in the Uruguay Round Agriculture Negotiations, Request for Exceptions for Import Restrictions as well as the Abolition of EC Export Subsidies), *Nihon Keizai Shimbun*, 27 December 1987, p. 2.

20 'Gatto no Shin Raundo, Nihon ga Teppai o Teian, Kajō Nōsanbutsu no Yushutsu Hojokin' (GATT Uruguay Round, Japan Proposes to Abolish Excess Agricultural Export Subsidies), *Asahi Shimbun*, 27 December 1987, p. 1.

21 The import restrictions on these two products were finally removed in late 1993 in Japan's offer to conclude the Uruguay Round negotiations. (GATT, *GATT Activities 1993*, 1994, p. 53.)

22 'Gyūniku · Orenji Kettyaku, Bei, Dotanba de Ayumiyori, Nihon no Mentsu Hairyo, "Nōgyō", Butai wa Shin Raundo e' (Beef and Orange Decision, United States, Last Minute Compromise, Consideration of Japan's Honor, 'Agriculture', The Scene Changes to the Uruguay Round), *Nihon Keizai Shimbun*, 20 June 1988, p. 3.

23 See 'Gyūniku · Orenji Jiyūka, Go Yatō, Shushō ni Hantai no Ketsugibun' (Beef and Orange Liberalization, Five Opposition Parties Present Opposing Written Resolution to the Prime Minister), *Nihon Keizai Shimbun*, 27 April 1988, p. 2.

24 'Takeshita Shushō Kaiken no Naiyō' (Contents of Interview with Prime Minister Takeshita), *Nihon Keizai Shimbun* (evening edition), 22 June 1988, p. 3.

25 'Jimin Shijiritsu, Nōrin Gyogyōsha de Kyūraku, Nōsanbutsu Jiyūka, Beika Hikisage Eikyō, 70% kara 59% ni, Shatō wa Baizō' (LDP Support Rate, Steep Decline among Farmers, Foresters and Fishermen, from 70% to 59%, Influenced by Liberalization of Agricultural Products, Reduction in Rice Price, Doubling in Support Rate for Socialist Party), *Mainichi Shimbun*, 8 September 1988, p. 2.

26 'Kome Jiyūka Hantai, Sanin de mo Ketsugi' (Against Rice Liberalization, Upper House Resolution as well), *Asahi Shimbun* (evening edition), 21 September 1988, p. 4; see also 'Kome no Jiyūka Hantai ni Kansuru Kokkai Ketsugi (Yōshi)' (Diet Resolution Concerning Opposition to Rice Liberalization (Main Points)), *Asahi Shimbun*, 15 June 1990, p. 3.

27 '"Kome Kaihō Mondai wa Atsukau beki de nai" Jimin Hakendanchō no Hatashi' ('The Rice Opening Problem Should not be Dealt With' says LDP Envoy Hata), *Asahi Shimbun*, 7 December 1988, p. 3.

28 'Nōgyō Hogo, Tanki de mo Sakugen, Kyūjūichinen Ikō wa Ōhaba ni, Gatto Jimu Reberu de Gōi' (Agricultural Protection, Reductions Even in the Short Term, Larger Reductions from 1991 on is the Agreement at the GATT Secretariat Level), *Asahi Shimbun*, 8 April 1989, p. 1.

29 'List of Representatives', GATT, Uruguay Round Trade Negotiations Committee, GATT Doc. No. MTN.TNC/INF/6/Rev.1, 6 April 1989, pp. 36–40.

30 'Trade Negotiations Committee Continues Discussions', *Japan Agrinfo Newsletter*, vol. 6, no. 10, June 1989, pp. 6–7, 7.

31 'Shushō Shoshin Hyōmei no Naiyō' (Content of Prime Minister's Speech Expressing his Beliefs), *Nihon Keizai Shimbun*, 6 June 1989, p. 7, as trans-

lated in 'Uno Vows to Lead Nation to Higher Standards, Better Relations' (Text of policy speech), *Japan Times*, 6 June 1989, p. 13.

32 'Jimin Kōkei Sōsai Erabi, Giin Sōkai de Tōhyō ga Yūryoku' (Selection of LDP Successor President, Likely by Vote in General Meeting of LDP Diet Members), *Asahi Shimbun*, 25 July 1989, p. 1.

33 'Nōson de no Jimin Zanpai, "Hogo" e Naiatsu Gaiatsu wa "Kaihō", Nōsei Jirenma Shinkokuka' (Crushing Defeat of the LDP by Farm Villages, Internal Pressure for 'Protectionism', External Pressure for 'Opening', Agricultural Administration Dilemma Becomes More Serious), *Asahi Shimbun*, 24 July 1989, p. 6.

34 'Agriculture . . . 25–26 September', GATT, *News of the Uruguay Round of Multilateral Trade Negotiations*, NUR 031, 16 October 1989, p. 8.

35 'Yunyū Shōheki Jūnen de Teppai o, Bei, Gatto Nōgyō Kōshō de Teian e, Kome mo Reigai Mitomezu, Yushutsu Hojokin wa "Gonen Inai"' (Abolish Import Barriers in Ten Years, the United States' Proposal to the GATT Agriculture Negotiations, Recognize No Exception for Rice, No Export Subsidies 'Within Five Years'), *Asahi Shimbun*, 22 October 1989, p. 9.

36 'Kyūjūnen Shūinsen Kōhosha Ankēto, Kome no Jiyūka, Hantai, Hachiwari o Koeru, Kome · Bōeki Fukinkō' (1990 Lower House Election Candidate Questionnaire, Opposition to Rice Liberalization more than Eighty per cent, Rice and the Trade Imbalance), *Mainichi Shimbun*, 10 February 1990, p. 17.

37 See 'Jimin, Tsuika Kōnin Fukume 284' (LDP, with Additional Nominations Includes 284), *Asahi Shimbun* (evening edition), 19 February 1990, p. 1.

38 See 'JSP Penetrates LDP-Dominant Rural Areas', *Mainichi Daily News*, 20 February 1990, p. 1.

39 See 'Shasetsu, Kome Ryūtsū no Issō no Jiyūka o' (Editorial, Concerning More Liberalization of Rice Distribution), *Mainichi Shimbun*, 30 April 1990, p. 5; see also Yamaji and Ito, 1993, pp. 351–3.

40 See 'Kome Sōdō, Kōmei "Eikyō Chōsa" Hatsugen no Hamon' (Rice Dispute, the Sensation caused by the Statements in Komeito's 'Influence Study'), *Asahi Shimbun*, 15 June 1990, p. 3.

41 'Framework Agreement on Agriculture Reform Programme, Draft Text by the Chairman', GATT, Uruguay Round Negotiating Group on Agriculture, GATT Doc. No. MTN.GNG/NG5/W/170, 11 July 1990, p. 4.

42 See 'Keizai Nyūsu Mondō, Kichō wa Bei no Kanzeikaan, Kome no Reigai Sochi o Shisa, Fujōshita Dozeu Teian' (Economic News Questions and Answers, the de Zeeuw Proposal Appears, Suggests Exceptional Measures for Rice, Basis is the American Tariffication Plan), *Asahi Shimbun*, 13 July 1990, p. 9.

43 'Kome "Kanzeika" Wakunai de, Tokubetsu Atsukai Nozomu Nihon, Tachiba Kurushiku, Gatto Nōgyō Kōshō Gichōan' (GATT Agriculture Negotiation Chair's Framework Agreement, Rice Within 'Tariffication' Framework, Japan Wants Special Treatment, Painful Position), *Mainichi Shimbun*, 4 July 1990, p. 9.

44 'Gatto Kōshō, "Nōgyō Kanzeika" ni Hantai, Nihon, Kitei Hōshin Kaezu' (GATT Negotiations, Opposing the 'Tariffication of Agriculture', Japan,

Doesn't Change Its Prearranged Plan), *Nihon Keizai Shimbun* (evening edition), 11 July 1990, p. 2.

45 See 'Kome Meguru Tachiba Gaikoku ni Setsumei o, Shushō, Nōsuishō ni Yōsei' (Prime Minister Requests Minister of Agriculture, Forestry and Fisheries to Explain Position Concerning Rice Abroad), *Asahi Shimbun*, 21 July 1990, p. 8.

46 See, for example, 'Nōsanbutsu "Kanzeika" Ukeire, Kome wa Wakugai Kuzusazu, Gatto Kōshō Seifu Hōshin, Kinkyūji Seigen Jyōken ni' (GATT Negotiations Government Plan, Accept the 'Tariffication' of Agricultural Products, Rice Outside the Framework Unchanged, Urgent Time for Conditions on Restrictions), *Nihon Keizai Shimbun*, 13 July 1990, p. 1.

47 'Kome Jikyū Kenji Kitai, Gatto Kōshō de Zenchū Kaichō' (Chairman of Central Union of Agricultural Co-operatives concerning GATT Negotiations, Expects to Hold Fast to Rice Self-Sufficiency Expectations), *Nihon Keizai Shimbun*, 28 July 1990, p. 5.

48 'Houston Economic Summit Economic Declaration, July 11, 1990', in Bush, *1990, Book II – July 1 to December 31, 1990*, 1991, p. 985.

49 'Trade Negotiations Committee, Chairman's summing-up at meeting of 26 July 1990', GATT, Uruguay Round Trade Negotiations Committee, GATT Doc. No. MTN.TNC/15, 30 July 1990, p. 3. See also 'Collective Determination to Succeed but Negotiations behind Schedule, Notes Dunkel at TNC', GATT, *News of the Uruguay Round of Multilateral Trade Negotiations*, NUR 039, 30 July 1990, p. 5.

50 See 'Kome Kaihō Fukumu Shinseisaku Ritsuan e Kōmei, Jūichigatsu made ni' (Towards a New Policy Plan from Komeito, including Rice Opening by November), *Asahi Shimbun*, 5 July 1990, p. 3.

51 'Kome Yunyū Waku, Jyūnengo ni Gojūman ton, Kōmei Seishinkaichō ga "Shiken" ' (Rice Import Framework, After Ten Years 500,000 Tons, Komeito Policy Chief's 'Personal View'), *Asahi Shimbun*, 11 August 1990, p. 2.

52 'Gov't Makes GATT Proposal', *Mainichi Daily News*, 29 September 1990, p. 7.

53 'Kome Kaihō no Eikyō, Nen Yonchō 8990 Oku En, Nōsuishō Shisan, Kanzen Jiyūka nara, GDP 1.5% Genshō' (Rice Opening Influence, 4 trillion, 899 billion Yen, Ministry of Agriculture, Forestry and Fisheries' Trial Calculation, if Complete Liberalization, 1.5% Decrease in GDP), *Mainichi Shimbun*, 28 October 1990, p. 3.

54 'Working Paper for Draft Agreement on Agriculture Trade Proposed by Uruguay Round Agriculture Negotiating Group Chairman Mats Hellstrom', *International Trade Reporter*, 12 December 1990, p. 1905.

55 See 'Saitei Yunyū Gimu, Kome Hakyū Kanōsei Mo' (Minimum Import Obligation, Possibility It Might Affect Rice As Well), *Asahi Shimbun* (evening edition), 7 December 1990, p. 1.

56 See, for example, 'Gyūniku · Orenji Raigetsu kara, Omowaku Nosete, Jiyūka Honban' (Beef and Oranges from Next Month, Expectations Pick Up, Liberalization About to Occur), *Nihon Keizai Shimbun*, 28 March 1991, p. 3.

57 'Kome Ichibu Jiyūka o Shuchō, ZenTsūsanshō, Bei de Kōen' (Former Minister of International Trade and Industry Insists on Partial Liberalization of the Rice Market during Lecture in the United States), *Asahi Shimbun* (evening edition), 27 April 1991, p. 2.

58 'Kanemarushi ga Kome Bubun Kaihōron, Toyama de Kōen, Shotoku Hoshō nado Jōken ni' (Kanemaru Speaks Concerning Partial Rice Opening Debate including Income Compensation and Other Matters as a Condition during Speech in Toyama), *Asahi Shimbun*, 20 May 1991, p. 1.

59 '"Kome Kaihō Ketsudan Hayaku", Jimin Sōmukaichō' ('Rice Opening Decision Soon', says LDP Policy Affairs Research Council Chairman), *Asahi Shimbun* (evening edition), 25 May 1991, p. 1.

60 'Bubun Jiyūka Taiō, Samitto ga Keiki, Kome de Takeshitashi' (Coping with Partial Rice Liberalization, Summit is an Opportunity, Mr Takeshita says), *Mainichi Shimbun*, 28 May 1991, p. 2.

61 'Shin Kakuryō ni Kiku, Tanabu Nōshō, "Kanzeika" Hantai o Tsuranuku' (Ask the New Cabinet Minister, Minister of Agriculture, Forestry and Fisheries Tanabu, Carrying Through Opposing 'Tariffication'), *Nihon Keizai Shimbun*, 9 November 1991, p. 5, trans. *Japan Agrinfo Newsletter*, vol. 9, no. 5, January 1992, pp. 8–9.

62 See 'Seiji Handan no Shōnenba, Shin Raundo Nōgyō Bunya Kyō Jikankyū Kaigō' (Uruguay Round Agriculture Group Meeting at Vice-Ministerial Level Today, Political Decision at the Crucial Moment), *Asahi Shimbun*, 20 November 1991, p. 9; 'Kome nado no Minimamu Akusesu, "Shonendo 3%" Gōi, Bei to EC nado, Bei Jiseki Daihyō Akasu' (Minimum Access for Rice and Other Products, 'First Year 3%' Agreement, the United States and the EC and Others, American Deputy USTR Reveals), *Asahi Shimbun*, 11 December 1991, p. 9; 'Nōgyō Bunya, Zentai Kaigō de Hantai o Kyōchō, Kanzeika ni Nihon' (Concerning Agriculture, Japan Emphasizes Opposition to Tariffication at General Meeting), *Asahi Shimbun*, 21 December 1991, p. 9.

63 'Shin Raundo Nōgyō Kaigō, Nihon · Kankoku nado Kanzeika Hantai Hyōmei' (Uruguay Round Agriculture Meeting, Japan, South Korea and others Express Opposition to Tariffication), *Asahi Shimbun*, 27 November 1991, p. 2.

64 'Paragraph 5, Part B, Agreement on Modalities for the Establishment of Specific Binding Commitments under the Reform Programme, L. Text on Agriculture', in 'Draft Final Act', GATT, Uruguay Round Trade Negotiations Committee, GATT Doc. No. MTN.TNC/W/FA, 20 December 1991, p. L.19.

65 'Jimin wa Hanpatsu, Kanjichō ga Danwa' (LDP Opposes, Talks to Chairman); 'Shokuryō Yunyūkoku o Mushi, Zenchū Kaichō' (Ignores Food Importing Countries, Central Union of Agricultural Co-operatives President says); 'Shakyōmin Kakutō Issei ni Hanpatsu' (Socialist, Communist, and Democratic Socialist Parties, Each Party Opposes Individually and Collectively), *Nihon Keizai Shimbun* (evening edition), 21 December 1991, p. 2.

66 'Raundo Seikō, Doryoku de Itti, Kankei Kakuryō Kondankai' (Round Successful, Effort to Agree, Concerned Cabinet Members' Informal Meeting), *Asahi Shimbun* (evening edition), 25 December 1991, p. 1.

67 '"Kuruma" de Taibei Kyōryoku o, Nentō Kaiken de Shushō ga Hyōmei, Kome, Nanraka no Sochi' (Work Together with the United States on 'Cars', Rice, Some Measures, Prime Minister States During New Year Interview), *Asahi Shimbun*, 1 January 1992, p. 2.

68 'Tōmen wa Kanzeika Kyohi Hōshin, Seifu · Jimintō, Beiō no Hannō o Chūshi' (For the Moment Plan to Refuse to Accept Tariffication, Government and the

LDP Closely Observe the United States' and EC's Reaction), *Asahi Shimbun* (evening edition), 21 December 1991, p. 1.

69   'Shin Raundo, Nōgyō de Seiji Ketsudan o, Keizai Dōyūkai ga Nentō Kenkai' (Uruguay Round, Make a Political Decision Concerning Agriculture, the Association of Corporate Executives' New Year Opinion), *Nihon Keizai Shimbun*, 4 January 1992, p. 3.

70   '"Kome" Ketsudan Sangatsu ni, Shushō Mitōshi, Kigen, Kakkoku mo Okure' (Prime Minister Forecasts 'Rice' Decision in March, Deadline, but Some Countries Are also Late), *Asahi Shimbun* (evening edition), 25 January 1992, p. 1.

71   'EC, Hogo Sakugen Furezu, Shin Raundo, Nōgyō Bunya de Kokubetsuhyō, Nihon, Kome nado Kūhaku' (Uruguay Round, Concerning Agriculture, Each Country's Schedule, EC Doesn't Touch Reductions in Protection, Japan Leaves Out Rice and Other Products), *Asahi Shimbun* (evening edition), 5 March 1992, p. 1.

72   See 'Taibei Hairyo de Kaishaku Henkō? Shokuryōchō, Beikokusan Kome Tenji Kyoka, Shokkanhō Unyō wa Tamamushiiro' (Food Agency Takes Into Consideration Changing Its Interpretation in Relation to the United States? Permission for Display of American-produced Rice, Equivocal Reply Concerning Application of Food Control Law), *Nihon Keizai Shimbun*, 2 April 1992, p. 5; 'Beikokusan Kome Tenji Mondai, Beigikai no Hanpatsu o Kenen, Hiruzu USTR Daihyō' (American Rice Agricultural Exhibit Problem, Worried about American Congressional Opposition, USTR Hills), *Nihon Keizai Shimbun* (evening edition), 19 March 1991, p. 2.

73   'Policies of Major Parties in Upper House Election', *Mainichi Daily News*, 24 July 1992, p. 2.

74   See 'Hosokawa ZenKumamoto Kenchiji, Saninsen de Kōhō Yōritsu, "Jiyū Shakai Rengō" Kessei e' (Previous Governor of Kumamoto Prefecture, Hosokawa, Supports Candidates in Upper House Election, Towards Formation of 'Liberal Socialist Union'), *Mainichi Shimbun* (evening edition), 7 May 1992, p. 1.

75   'Trade Negotiations Committee, Twenty-Eighth Meeting: 28 July 1993', GATT, Uruguay Round Trade Negotiations Committee, GATT Doc. No. MTN.TNC/32, 18 August 1993, pp. 1–2, 9.

76   'Kome Yunyū 100 Man Ton Ijō ni, Sakkyō Shisū, Sara ni Akka Mitōshi' (Rice Imports of More than One Million Tons, Harvest Index, Forecast Getting Even Worse), *Asahi Shimbun*, 30 September 1993, p. 1.

77   'Shushō, "Ketsudan" ni nao Kabe, Shin Raundo, Yoron no Henka Niramu' (Uruguay Round, Prime Minister, 'Decision' Still Not Easy, Keeping an Eye on Changes in Public Opinion), *Asahi Shimbun*, 1 October 1993, p. 2.

78   'Shōhizeiritsu, Jieitai Hō, Kome Kaihō, Renritsu Yusaburu Seisaku Chōsei, Teikō Kuzusanu Shatō, Shijisō nirami Kakutō Jijō mo' (Consumption Tax Rate, Self-Defense Law, Rice Opening, Coalition-Shaking Policy Adjustments, Do Not Destroy Opposition of Socialist Party, a Look at the Support Layer as well as the Circumstances of Each Party), *Mainichi Shimbun*, 3 October 1993, p. 2.

79   'Kome Kanzeika Ukeire Dashin, Seifu, Jisshi, Rokunen Hodo Yūyo, Mazu 4.5% Teido Yunyū ka' (Government Explores Accepting Rice Tariffication,

if Put into Effect, About 6 Years Postponement, at first about 4.5% Imported?), *Asahi Shimbun* (evening edition), 15 October 1993, p. 1.

80 'Shushō, Jijitsujō Ukeire, Kome Bubun Kaihō no Chōseian' (Prime Minister Receives the Facts, Partial Rice Opening Compromise Plan), *Asahi Shimbun* (evening edition), 7 December 1993, p. 1.

81 'Shatō Sanyaku ga Hantai o Kakunin' (Three Key Officials of the Socialist Party Confirm their Opposition), *Asahi Shimbun* (evening edition), 7 December 1993, p. 1.

82 'Nayameru Shatō Meguri Tsunahiki, Kome Bubun Kaihō, Nōgyō Dantai "Saigo no Tanomi", Kantei wa Settoku e Denwa Kōsei' (Worried Socialist Party Turns to Tug of War, Rice Partial Opening, Agricultural Groups' 'Last Request' Phone Offensive to Persuade the Government), *Asahi Shimbun*, 8 December 1993, p. 27.

83 'Nōgyō "Chōseian" o Seishiki Teiji, Dunī Gichō, Kome, "Kanzeika Yūyo" Moru, Shichinenme ikō Tsuika Jōho Hitsuyō ni' (Concerning Rice, Chairman Denis Formally Presents Agriculture 'Compromise Plan', Includes 'Tariffication Postponement', During and After Seventh Year Additional Concessions Are Necessary), *Mainichi Shimbun* (evening edition), 9 December 1993, p. 1.

84 ' "Hantai da ga Ryō to suru", Shatō Iinchō, Kunō no Kaiken' ('Opposed, but Understand', Socialist Party Chairman, Distressed Interview), *Asahi Shimbun* (evening edition), 14 December 1993, p. 11.

85 'Shushō "Kokueki Kangae Ketsudan", Kome Bubun Kaihō, Kaiken no Yōshi' (Prime Minister's 'Decision Thinking of Country's Benefit', Partial Rice Opening, Highlights of Press Conference), *Asahi Shimbun* (evening edition), 14 December 1993, p. 3.

86 'Keizaikai, "Nihon no Sekinin" to Hyōka, Sōkyū na Nōsei Tatenaoshi Yōkyū' (The Economic World, 'Japan's Responsibility' and Judgment, Request for Urgent Agricultural Policy Rebuilding), *Mainichi Shimbun* (evening edition), 14 December 1993, p. 3

87 'Farmers Denounce Gov't Decision', *Mainichi Daily News*, 15 December 1993, p. 1.

88 See 'Chōseian no "Tsuika Jōho", Yotōnai Chōsei no Shōten ni, Kome Kanzeika Tokurei Sochi Enchō' (Rice Tariffication, Extended Measures for Special Case, 'Additional Concessions' in the Compromise Plan, Parties in Power Focus on Adjustment), *Mainichi Shimbun*, 9 December 1993, p. 3.

89 'The Final Act of the Uruguay Round: Press Summary', in GATT, *News of the Uruguay Round of Multilateral Trade Negotiations*, NUR 080, 14 December 1993, p. 9.

90 'Farm Products to See Tariffs of up to 564%', *Mainichi Daily News*, 18 December 1993, p. 7. These figures can be found in the Schedule for Japan in GATT, 1994a, Vol. 11, p. 9370.

91 'Paragraph 1, Section A of Annex 5, Special Treatment with Respect to Paragraph 2 of Article 4', Agreement on Agriculture in GATT, 1994b, p. 65.

92 'Hata Seiken Shūake Hossoku e' (Hata Administration to Start at the Beginning of the Week), *Asahi Shimbun* (evening edition), 22 April 1994, p. 1.

93 'Murayama Shushō ga Tanjō' (Murayama becomes Prime Minister), *Asahi Shimbun*, 30 June 1994, p. 1.

94   'WTO Kyōtei Shōnin, "Nōgyō Rokuchō En" ni Umami to Tomadoi' (WTO Agreement Approval, '6 trillion Yen' to Agriculture, Advantages and Not Sure What to Do') *Asahi Shimbun*, 9 December 1994, p. 3; see also 'WTO Taisaku Motomeru Ketsugi, Shūin de Saitaku' (Resolution Requesting WTO Measures, Adoption by the Lower House), *Asahi Shimbun*, 3 December 1994, p. 2.
95   Yuji Iwasawa, 1997, p. 173. Concerning the effects of implementation of the Uruguay Round Agreement on Agriculture on Japanese agricultural policy, see Naraomi Imamura, *et al.* 1997; Shokuryō Seisaku Kenkyūkai, 1999.
96   Japan, Ministry of Agriculture, Forestry and Fisheries, 1987, p. 14.
97   Hisane Masaki, 'GATT hopes threaten Miyazawa', *Japan Times*, 20 October 1992, p. 1.
98   For a description of this coalition and the pro-liberalization coalition, but which does not discuss their evolution during the negotiations, see Rapkin and George, 1993, pp. 60–71.
99   See, for example, 'Hantai Demo ga Kokkai Shikichi ni' (Protest Demonstration in the Japanese Diet Site), *Asahi Shimbun*, 10 December 1993, p. 1.

5   US–Japan Trade Policymaking during the Uruguay Round Agriculture Negotiations

1   'Remarks of the President and Prime Minister Yasuhiro Nakasone of Japan Following Their Meetings in Tokyo, November 10, 1983', in Reagan, *1983, Book II – July 2 to December 31, 1983*, 1985, pp. 1568–9.
2   See 'GATT Agricultural Committee Approves First Draft of Proposals to Reduce Protectionism', *International Trade Reporter's U.S. Import Weekly*, 20 June 1984, p. 1143.
3   'Shin Raundo Junbi Utau, Gutaiteki Tejun nao Chōsei, Samitto Keizai Sengen Katamaru, Zaisei Akaji o Sakugen, Bei Nazashi no Yōkyū Sakeru' (Uruguay Round Preparations Stated, Concrete Process Still Being Adjusted, Summit Economic Declaration Is Being Finalized, Budget Deficits to be Reduced, Demands on the United States are Avoided), *Asahi Shimbun* (evening edition), 9 June 1984, p. 1.
4   'London Economic Summit Conference Declaration, June 9, 1984', in Reagan, *1984, Book I – January 1 to June 29, 1984*, 1986, p. 832.
5   'Statement of Daniel G. Amstutz, Under Secretary, International Affairs and Commodity Programs, U.S. Department of Agriculture', in US Congress, House Committee on Agriculture, *Agricultural Provision Proposals to Omnibus Trade Legislation: Hearing*, 99th Cong., 2nd sess., 15 April 1986, pp. 5–7,6.
6   'Sengenan, Nihon Shudō de, "Shin Raundo" Chiba Junēbu Taishi Kataru, Junbi Sagyō, Yama wa Korekara' (Declaration Proposal, Under Japan's Leadership, Japan's Ambassador to the 'Uruguay Round' in Geneva [Kazuo] Chiba Talks, Preparatory Work, Peak From Now On), *Nihon Keizai Shimbun*, 24 April 1986, p. 3.
7   'OECD Ministers Agree on Comprehensive Agenda for New Round of GATT Trade Talks', *International Trade Reporter*, 23 April 1986, pp. 532–3.
8   'U.S. Trade Representative Says Substantial Agreement Reached on GATT Round Agenda', *International Trade Reporter*, 4 June 1986, pp. 736–7; 'Shin

Raundo Kaishi e Zenshin, Kokusai Bōeki Kaigi ga Heimaku' (Uruguay Round Beginning Advances, Drawing the Curtain on International Trade Conference), *Asahi Shimbun*, 2 June 1986, p. 2.

9   'EC Members Agree Services, Farm Subsidies Should be on Agenda for New GATT MTN Round', *International Trade Reporter*, 25 June 1986, p. 824.

10   Paul Lewis, 'Aims on Trade Talks Outlined', *New York Times*, 19 July 1986, p. 40.

11   'Preparatory Committee Fails to Agree on MTN Agenda, GATT Ministers Meeting Will Tackle', *International Trade Reporter*, 6 August 1986, pp. 1002–3, p. 1002.

12   'Yonkyoku Tsūshō Kaigi Oe Kyōdō Kaiken, Gatto no Shin Raundo, Kōshō Kaishi e Kyōryoku Gōi, Nōsanbutsu Yushutsu Hojokin nao Kuichigai' (Quadrilateral Trade Ministers Conference Ends, Joint Press Conference, GATT Uruguay Round, Negotiations Start Towards Cooperative Agreement, Still Differences Concerning Agricultural Export Subsidies), *Asahi Shimbun*, 8 September 1986, p. 2.

13   'Broad MTN Agenda Agreed to in Quadrilateral Meeting, But Differences Remain on Issues', *International Trade Reporter*, 10 September 1986, p. 1006.

14   'Nōgyō Bōeki Chōsei ni Jishin, Gatto Nōgyō Buchō, Shin Raundo e Kenkai' (GATT Agriculture Division Chief, Confidence Concerning Agricultural Trade Adjustment, Views Concerning the Uruguay Round), *Nihon Keizai Shimbun*, 5 September 1986, p. 7.

15   'Kakkoku Shin Raundo ni Omowaku Fukuzatsu, Shin Bunya · Nōsanbutsu ga Shōten' (Each Country's Complicated Intentions Concerning the Uruguay Round, Focus on New Areas and Agricultural Products), *Asahi Shimbun*, 11 September 1986, p. 4.

16   'Response of Allan I. Mendelowitz, Senior Associate Director, National Security and International Affairs Division, U.S. General Accounting Office', in US Congress, House Committee on Agriculture, Subcommittee on Department Operations, Research, and Foreign Agriculture, *Hearing to Receive GAO Reports on U.S. Department of Agriculture's Agricultural Trade Programs; and Summary of Trade Sessions Held in Uruguay*, 99th Cong., 2nd sess., 29 September 1986, p. 43.

17   See 'Battle Lines Drawn at GATT Talks – Discussion of Services, Farm Subsidies Among Issues', *Mainichi Daily News*, 17 September 1986, p. 6.

18   'GATT Launches Uruguay Round as Consensus Reached on Services, Agricultural Trade', *International Trade Reporter*, 24 September 1986, pp. 1150–2, 1150.

19   'Punta del Este Declaration, Ministerial Declaration of 20 September 1986', in GATT, *GATT Activities 1986*, 1987, p. 21.

20   'Kome Jiyūka wa Muri, Gaishō, Bei Tsūshō Daihyō ni Genmei' (Rice Liberalization Impossible, Foreign Minister Declares to the USTR), *Asahi Shimbun* (evening edition), 17 September 1986, p. 2.

21   'List of Representatives – Group of Negotiations on Goods – Group of Negotiations on Services', GATT, Uruguay Round Trade Negotiations Committee, GATT Doc. No. MTN.TNC/INF/1, MTN.GNG/INF/1, 27 October 1986, pp. 15–16, 24.

22 'Fifth Meeting of the Group of Negotiations on Goods: Record of Decisions Taken', GATT, Uruguay Round Group of Negotiations on Goods, GATT Doc. No. MTN.GNG/5, 9 February 1987, pp. 3–4, 10–11.

23 'Chairmen of Negotiating Groups', GATT, Uruguay Round Group of Negotiations on Goods, GATT Doc. No. MTN.GNG/W/9, 12 February 1987.

24 A list of derestricted and restricted documents related to these negotiations is available in 'Derestriction of Uruguay Round Documents: Documents proposed for derestriction on 3 January 1997', GATT, Uruguay Round Trade Negotiations Committee, GATT Doc. No. MTN.TNC/46, 31 October 1996.

25 'Annex: Attendance of International Organizations in the Proceedings of the Uruguay Round, Adopted by the Trade Negotiations Committee, 3 July 1987, Trade Negotiations Committee, Third Meeting: 3 July 1987', GATT, Uruguay Round Trade Negotiations Committee, GATT Doc. No. MTN.TNC/3, 22 July 1987, pp. 5–6.

26 See GATT, *Uruguay Round Consolidated List and Index of Documents Issued between 1986 and 1994* (prepared for microfiche users only) (Geneva: General Agreement on Tariffs and Trade, 1995–2000).

27 'Summary of Major Problems and their Causes as Identified Thus Far and of Issues Considered Relevant, Note by the Secretariat', GATT, Uruguay Round Negotiating Group on Agriculture, GATT Doc. No. MTN.GNG/NG5/W/2, 31 March 1987, p. 3.

28 Hisane Masaki and Ken Moritsugu, 'Trading Partners Urge Quick Action to Reduce Surplus', *Japan Times*, 26 April 1987, p. 1.

29 'Nijūyokka kara Mie de Yonkyoku Tsūshō Kaigi, Bōeki Fukinkō, Shin Raundo, Nidai Tēma' (Starting April 24th in Mie, A Quadrilateral Meeting, Two Large Themes, Trade Imbalance and the Uruguay Round), *Asahi Shimbun*, 22 April 1987, p. 9.

30 'U.S.–Japanese Agricultural Talks Conclude with U.S. Official Voicing Disappointment', *International Trade Reporter*, 22 April 1987, p. 536.

31 'Kome Kaihō, Nikoku Kyōgi mo Yōkyū, Bei Nōmuchōkan, Nōsuishō Kobami Monowakare' (Rice Opening, U.S. Agriculture Secretary Makes Demands in Bilateral Conference as well, Minister of Agriculture, Forestry and Fisheries Refuses, Fail to Reach Agreement), *Asahi Shimbun*, 21 April 1987, p. 1.

32 'U.S. Officials Meet with Agriculture Minister', *Japan Agrinfo Newsletter*, vol. 4, no. 10, May 1987, pp. 4–5.

33 'Lack of Progress in U.S.–Japan Talks Leads U.S. Officials to Repeat Trade Bill Warnings', *International Trade Reporter*, 29 April 1987, p. 576.

34 'Written Responses to Questions Submitted by the Japanese Newspaper *Asahi Shimbun*, 28 April 1987', in Reagan, *1987, Book I – January 1 to July 3, 1987*, 1989, p. 430.

35 'Summary of the Main Points Raised in the Course of the Group's Consideration of Basic Principles to Govern World Trade in Agriculture: 5–6 May 1987, Note by the Secretariat', GATT, Uruguay Round Negotiating Group on Agriculture, GATT Doc. No. MTN.GNG/NG5/W/12, 22 June 1987, pp. 8, 11.

36 Ibid., pp. 1–2.

37 Ken Moritsugu, 'Venice Summit: Valuable Chance to Solve Key Issues: Diplomat [Kazuo Chiba] Says Multilateral Answers are Needed to World Trade Problems', *Japan Times*, 5 June 1987, p. 1.

38 'Text of Economic Declaration by 7 Leading Industrial Nations', *New York Times*, 11 June 1987, sec. A, p. 16.

39 'OECD Ministers Meeting Ends with Agreement on Farm Reform and Economic Reaffirmations', *International Trade Reporter*, 20 May 1987, p. 666.

40 'OECD Report Shows Japanese Farm Subsidies to be Largest, Reaching Up to 100 Percent', *International Trade Reporter*, 20 May 1987, pp. 669–70, 669; 'Summary of Studies on Problems Affecting Trade in Agriculture and their Causes, Note by the Secretariat, Addendum, OECD: National Policies and Agricultural Trade', GATT, Uruguay Round Negotiating Group on Agriculture, GATT Doc. No. MTN.GNG/NG5/W/3/Add.1, 25 June 1987.

41 'United States Proposal of 6 July 1987 for Negotiations on Agriculture', in Petersmann and Hilf, 1988, p. 585.

42 See 'Major Farming Nations Applaud U.S. Subsidy Reform Plan, Yet Question Its Feasibility', *International Trade Reporter*, 8 July 1987, pp. 861–2, 861.

43 'Yunyū Seigen Yōnin o, Nihon "Jōken Tsuki" Shuchō, Gatto Kōshō' (GATT Negotiations, Japan Insists on 'Conditions' in Order to Accept Import Restriction Provisions), *Asahi Shimbun* (evening edition), 7 July 1987, p. 1.

44 'Summary of Main Points Raised at the Fourth Meeting of the Negotiating Group on Agriculture (26–27 October 1987), Note by the Secretariat', GATT, Uruguay Round Negotiating Group on Agriculture, GATT Doc. No. MTN.GNG/NG5/W/33, 19 November 1987.

45 GATT, *GATT Activities 1987*, 1988, p. 31.

46 Ibid.

47 'Summary of Main Points Raised at the Sixth Meeting of the Negotiating Group on Agriculture (15–17 February 1988), Note by the Secretariat', GATT, Uruguay Round Negotiating Group on Agriculture, GATT Doc. No. MTN.GNG/NG5/W/52, 17 March 1988, p. 1.

48 See 'EC Rejects U.S. Attempt to Speed up GATT Negotiations, Says U.S. "Ignoring Reality"', *International Trade Reporter*, 24 February 1988, pp. 251–2.

49 'Summary of Main Points Raised at the Eighth Meeting of the Negotiating Group on Agriculture (9–10 June 1988), Note by the Secretariat', GATT, Uruguay Round Negotiating Group on Agriculture, GATT Doc. No. MTN.GNG/NG5/W/63, 28 June 1988.

50 'Agriculture . . . 9 and 10 June', GATT, *News of the Uruguay Round of Multilateral Trade Negotiations*, NUR 017, 30 June 1988, p. 7.

51 'Agriculture Discussed at Summit', *Japan Agrinfo Newsletter*, vol. 6, no. 1, August 1988, pp. 2–3, 3; see also 'Chūchōki de Sakugen Doryoku, Nōgyō Hojokin, "Hogo Shihyō" o Dōnyū' (Agricultural Subsidies, Medium- to Long-Term Reduction Efforts, 'Indices of Protection' Introduced), *Nihon Keizai Shimbun*, 22 June 1988, p. 1.

52 'Toronto Economic Summit Conference Economic Declaration, June 21, 1988', in Reagan, *1988, Book I – January 1 to July 1, 1988*, 1990, p. 800.

53 'Agriculture . . . 13 and 14 July', GATT, *News of the Uruguay Round of Multilateral Trade Negotiations*, NUR 018, 2 August 1988, pp. 9–10.

54 'Summary of Main Points Raised at the Ninth Meeting of the Negotiating Group on Agriculture (13–14 July 1988), Note by the Secretariat', GATT, Uruguay Round Negotiating Group on Agriculture, GATT Doc. No. MTN.GNG/NG5/W/73, 29 July 1988, p. 5.

55 'Group of Negotiations on Goods, Eleventh Meeting: 25 and 26 July 1988', GATT, Uruguay Round Group of Negotiations on Goods, GATT Doc. No. MTN.GNG/12, 15 August 1988, p. 8.

56 'Agriculture . . . 12 and 13 September', GATT, *News of the Uruguay Round of Multilateral Trade Negotiations*, NUR 019, 5 October 1988, pp. 2/3–4.

57 'Summary of Main Points Raised at the Tenth Meeting of the Negotiating Group on Agriculture (12–13 September 1988), Note by the Secretariat', GATT, Uruguay Round Negotiating Group on Agriculture, GATT Doc. No. MTN.GNG/NG5/W/79, 29 September 1988, p. 1.

58 'U.S. Negotiator Reports Some Progress in GATT Farm Talks, Calls EC Obstacle', *International Trade Reporter*, 21 September 1988, p. 1271.

59 'Summary of Main Points Raised at the Eleventh Meeting of the Negotiating Group on Agriculture (13–14 October 1988), Note by the Secretariat', GATT, Uruguay Round Negotiating Group on Agriculture, GATT Doc. No. MTN.GNG/NG5/W/86, 10 November 1988, pp. 3–4.

60 Ibid., p. 3.

61 'Trade in Agricultural Products . . . 11–14 October', GATT, *News of the Uruguay Round of Multilateral Trade Negotiations*, NUR 020, 4 November 1988, p. 7.

62 'Meeting at Ministerial Level, December 1988: Checklist of documents, Revision', GATT, Uruguay Round Trade Negotiations Committee, GATT Doc. No. MTN.TNC/W/13/Rev.1, 1 December 1988, p. 1.

63 'Report by the Chairman of the Negotiating Group on Agriculture, Part B – Points for Decision', GATT, Uruguay Round Group of Negotiations on Goods, GATT Doc. No. MTN.GNG/16, 30 November 1988, p. 2. See also 'Trade Negotiations Committee Meeting at Ministerial Level, Montreal, December 1988', GATT, Uruguay Round Trade Negotiations Committee, GATT Doc. No. MTN.TNC/7(MIN), 9 December 1988, p. 11.

64 'Remarks to the National Chamber Foundation, November 17, 1988', in Reagan, *1988–89, Book II – July 2, 1988 to January 19, 1989*, 1991, p. 1523.

65 Kyodo, ' "Realistic" Look at Farm Trade Reform, Uno Urges', *Japan Times*, 7 December 1988, p. 1. See also 'Kome Kaihō wa Kobamu Shisei, Uno Gaishō Enzetsu e "Shokuryō Anpō" Zenmen ni' (Stance to Refuse Rice Opening, in Foreign Minister Uno's Speech 'Food Security' in Front), *Asahi Shimbun*, 6 December 1988, p. 3.

66 'Progress at Montreal Midterm Should Keep MTN Momentum Going, Japanese Officials Say', *International Trade Reporter*, 14 December 1988, pp. 1629–30.

67 'Trade Negotiations Committee Meeting at Ministerial Level, Montreal, December 1988', GATT, Uruguay Round Trade Negotiations Committee, GATT Doc. No. MTN.TNC/7(MIN), 9 December 1988, pp. 1, 10.

68 'Meeting at Ministerial Level, Palais des Congrès, Montreal (Canada), 5–9 December 1988', GATT, Uruguay Round Trade Negotiations Committee, GATT Doc. No. MTN.TNC/8(MIN), 17 January 1989, p. 15.

69 ' "Seika Atta", Tamura Tsūsanshō' ('Achieved Results', Minister of International Trade and Industry Tamura Says), *Mainichi Shimbun*, 10 December 1988, p. 9.
70 See 'Agreement on says Agriculture Within Reach as Uruguay Round Review Set to Resume', *International Trade Reporter*, 5 April 1989, pp. 410–12.
71 'Trade Negotiations Committee Meeting at Level of High Officials, Geneva, 5–8 April 1989', GATT, Uruguay Round Trade Negotiations Committee, GATT Doc. No. MTN.TNC/9, 11 April 1989, pp. 3–6.
72 'Nōgyō Bunya Gatto Kōshō, Hogo Suijun Sakugen Moru Kōsan, Tanki Sochi de Bei · EC Dōchō' (GATT Agriculture Negotiations, Greater Probability of Reduction in Levels of Protection, the United States and EC Agree to Short-Term Measures), *Asahi Shimbun* (evening edition), 7 April 1989, p. 1.
73 'Nōgyō Hogo, Tanki de mo Sakugen, Kyūjūichinen Ikō wa Ōhaba ni Gatto Jimu Reberu de Gōi' (Agricultural Protection, Reductions Even in the Short Term, Larger Reductions from 1991 On is the Agreement at the GATT Secretariat Level), *Asahi Shimbun*, 8 April 1989, p. 1.
74 'Trade Negotiations Committee, Ninth Meeting: 27 July 1989', GATT, Uruguay Round Trade Negotiations Committee, GATT Doc. No. MTN.TNC/12, 16 August 1989, p. 5. See also Dunkel, 1990, p. 9.
75 'GATT Rules and Disciplines Relating to Agriculture, Note by the Secretariat', GATT, Uruguay Round Negotiating Group on Agriculture, GATT Doc. No. MTN.GNG/NG5/W/95, 4 July 1989.
76 'Summary of the Main Points Raised at the Fourteenth Meeting of the Negotiating Group on Agriculture (10–12 July 1989), Note by the Secretariat', GATT, Uruguay Round Negotiating Group on Agriculture, GATT Doc. No. MTN.GNG/NG5/W/103, 4 September 1989, p. 3.
77 'Agriculture . . . 10 to 12 July', GATT, *News of the Uruguay Round of Multilateral Trade Negotiations*, NUR 030, 3 August 1989, p. 8.
78 'Summary of the Main Points Raised at the Fourteenth Meeting of the Negotiating Group on Agriculture (10–12 July 1989), Note by the Secretariat', GATT, Uruguay Round Negotiating Group on Agriculture, GATT Doc. No. MTN.GNG/NG5/W/103, 4 September 1989, p. 3.
79 'Summary of the Main Points Raised at the Fifteenth Meeting of the Negotiating Group on Agriculture (25–26 September 1989) Note by the Secretariat', GATT, Uruguay Round Negotiating Group on Agriculture, GATT Doc. No. MTN.GNG/NG5/W/113, 23 October 1989, p. 4.
80 'Agriculture . . . 25–26 September', GATT, *News of the Uruguay Round of Multilateral Trade Negotiations*, NUR 031, 16 October 1989, p. 8.
81 'Summary of the Main Points Raised at the Sixteenth Meeting of the Negotiating Group on Agriculture (25–26 October 1989) Note by the Secretariat', GATT, Uruguay Round Negotiating Group on Agriculture, GATT Doc. No. MTN.GNG/NG5/W/123, 10 November 1989, p. 3.
82 'Agriculture . . . 25–26 October', GATT, *News of the Uruguay Round of Multilateral Trade Negotiations*, NUR 032, 21 November 1989, p. 6.
83 'Jitsugensei ni Gimon, Nōsuishō Shiteki' (Doubts about Putting into Practice, the Ministry of Agriculture, Forestry and Fisheries Points Out), *Nihon Keizai Shimbun* (evening edition), 25 October 1989, p. 1.
84 'Summary of the Main Points Raised at the Seventeenth Meeting of the Negotiating Group on Agriculture (27–28 November 1989) Note by

the Secretariat', GATT, Uruguay Round Negotiating Group on Agriculture, GATT Doc. No. MTN.GNG/NG5/W/139, 14 December 1989, pp. 1–4.

85 '"Kome Jikyū Kakuho" ni Nerai, Gatto Nōgyō Kōshō, Nihon Teian no Zenyō Hanmei' (GATT Agriculture Negotiations, Aim is 'Securing Rice Self-Sufficiency', Japan's Proposal Makes Clear the Whole Picture), *Asahi Shimbun* (evening edition), 21 November 1989, p. 1.

86 'Summary of the Main Points Raised at the Seventeenth Meeting of the Negotiating Group on Agriculture (27–28 November 1989) Note by the Secretariat', GATT, Uruguay Round Negotiating Group on Agriculture, GATT Doc. No. MTN.GNG/NG5/W/139, 14 December 1989, p. 5.

87 'Summary of the Main Points Raised at the Eighteenth Meeting of the Negotiating Group on Agriculture (19–20 December 1989) Note by the Secretariat', GATT, Uruguay Round Negotiating Group on Agriculture, GATT Doc. No. MTN.GNG/NG5/W/149, 6 February 1990, pp. 4–7.

88 Ibid., pp. 1–3.

89 Ibid., pp. 3–4.

90 'Japanese Agriculture Official Says Election Will Not Affect Position on Food Security', *International Trade Reporter*, 10 January 1990, p. 49.

91 'Farm Organizations Back Agriculture Reform Through Strengthening GATT Article XI Rules', *International Trade Reporter*, 14 February 1990, pp. 231–2; see also Naokazu Takeuchi, Consumers Union of Japan, 'Jidai no Me, Kome Kaihō, Shōhisha Yōgo ni arazu, Shokuryō Kiki no Ninshiki ga Hitsuyō, Bei Shuchō wa Kigyō Rieki Daiben' (Eye of the Times, Rice Opening, It Isn't to Protect Consumers, Necessary to Be Aware of Food Crisis, American Insistence Speaks for Business Interests), *Mainichi Shimbun*, 7 April 1990, p. 4, translated in 'Demands to Open Rice Market Benefiting Only Multinationals', *Japan Agrinfo Newsletter*, vol. 7, no. 11, July 1990, pp. 11–13.

92 'Synoptic Table of Negotiating Proposals Submitted Pursuant to Paragraph 11 of the Mid-Term Review Agreement on Agriculture, Note by the Secretariat', GATT, Uruguay Round Negotiating Group on Agriculture, GATT Doc. No. MTN.GNG/NG5/W/150, 12 February 1990, pp. 10–18.

93 'Synoptic Table of Negotiating Proposals Submitted Pursuant to Paragraph 11 of the Mid-Term Review Agreement on Agriculture, Note by the Secretariat, Revision', GATT, Uruguay Round Negotiating Group on Agriculture, GATT Doc. No. MTN.GNG/NG5/W/150/Rev.1, 2 April 1990, pp. 13–26.

94 'Clarification and Elaboration of Elements of Detailed Proposals Submitted pursuant to the Mid-Term Review Decision, Note by the Secretariat', GATT, Uruguay Round Negotiating Group on Agriculture, GATT Doc. No. MTN.GNG/NG5/W/161, 4 April 1990, pp. 43–62.

95 'Clarification and Elaboration of Elements of Detailed Proposals Submitted Pursuant to the Mid-Term Review Decision, Note by the Secretariat, Addendum', GATT, Uruguay Round Negotiating Group on Agriculture, GATT Doc. No. MTN.GNG/NG5/W/161/Add.2, 18 May 1990.

96 'Group of Negotiations on Goods, Sixteenth meeting: 9 April 1990', GATT, Uruguay Round Group of Negotiations on Goods, GATT Doc. No. MTN.GNG/22, 8 May 1990, p. 3.

97 'Nōgyō Kaikaku ni Doryoku o, Gatto Jimukyokuchō, Yamamoto Nōshō ni Yōsei' (GATT Director-General Requests Minister of Agriculture, Forestry

and Fisheries Yamamoto to Make an Effort in Agricultural Reform), *Nihon Keizai Shimbun*, 3 June 1990, p. 3.

98 'Japan's Agriculture Minister Expects Problem in Meeting July Deadline for GATT Agreement', *International Trade Reporter*, 27 June 1990, p. 967.

99 'Framework Agreement on Agriculture Reform Programme, Draft Text by the Chairman', GATT, Uruguay Round Negotiating Group on Agriculture, GATT Doc. No. MTN.GNG/NG5/W/170, 11 July 1990, pp. 1, 2, 4.

100 'Santiago Meeting of Cairns Group Ministers: 4–6 July 1990, Press Communiqué and Conclusions, Submitted by Australia on Behalf of the Cairns Group', GATT, Uruguay Round Negotiating Group on Agriculture, GATT Doc. No. MTN.GNG/NG5/W/175, 24 July 1990.

101 'Houston Economic Summit Economic Declaration, July 11, 1990', in Bush, *1990, Book II – July 1 to December 31, 1990*, 1991, p. 985.

102 See 'Kome Meguru Tachiba, Gaikoku ni Setsumei o, Shushō, Nōsuishō ni Yōsei' (Prime Minister Requests Minister of Agriculture, Forestry and Fisheries to Explain Position Concerning Rice Abroad), *Asahi Shimbun*, 21 July 1990, p. 8.

103 'Houston Economic Summit Economic Declaration, July 11, 1990', in Bush, *1990, Book II – July 1 to December 31, 1990*, pp. 984–5.

104 'Trade Negotiations Committee, Chairman's summing-up at meeting of 26 July 1990', GATT, Uruguay Round Trade Negotiations Committee, GATT Doc. No. MTN.TNC/15, 30 July 1990, p. 3.

105 'Agriculture . . . 27–29 August', GATT, *News of the Uruguay Round of Multilateral Trade Negotiations*, NUR 041, 9 October 1990, p. 2/3.

106 'Nōgyō Hogo 30% Sakugen, Seifu, Gatto Kōshō de Teian e' (Government Proposes at GATT Negotiations 30% Reductions in Agricultural Protection), *Nihon Keizai Shimbun*, 27 September 1990, p. 1.

107 'Japan Will Offer 30 Percent Subsidy Cuts over 10 Years at Uruguay Round Negotiations', *International Trade Reporter*, 3 October 1990, pp. 1499–1500.

108 'Twenty-Fifth Session of the Negotiating Group on Agriculture, Note by the Chairman', GATT, Uruguay Round Negotiating Group on Agriculture, GATT Doc. No. MTN.GNG/NG5/26, 4 October 1990, p. 1.

109 'Bei, Nōgyō Mondai de Fuman, Kanzeika Toriirezu Shitsubō, Nichibei Bōeki Iinkai' (US–Japan Trade Committee, the United States is Not Satisfied With the Agricultural Problem, Disappointed With Rejection of Tariffication), *Mainichi Shimbun* (evening edition), 3 October 1990, p. 2.

110 'U.S. Advances Final Farm Proposal Calling for 90 Percent Cuts in Export Subsidy Levels', *International Trade Reporter*, 17 October 1990, pp. 1564–5. See also 'U.S. Proposal for Agricultural Negotiations in the Uruguay Round of Multilateral Trade Talks with Accompanying Fact Sheet Released by the Office of the U.S. Trade Representative Oct. 15, 1990', *International Trade Reporter*, 17 October 1990, pp. 1594–9.

111 'Twenty-Fifth (Resumed) Session of the Negotiating Group on Agriculture, Note by the Chairman, Revision', GATT, Uruguay Round Negotiating Group on Agriculture, GATT Doc. No. MTN.GNG/NG5/27/Rev.1, 29 October 1990.

112 'Appendix B, Survey of Offers' in 'Agriculture' in 'Draft Final Act Embodying the Results of the Uruguay Round of Multilateral Trade Negotiations,

Revision', GATT, Uruguay Round Trade Negotiations Committee, GATT Doc. No. MTN.TNC/W/35/Rev.1, 3 December 1990, pp. 139–50.

113   'U.S., Others Blame EC for Failure in Brussels to Agree on New Rules to Govern World Trade', *International Trade Reporter*, 12 December 1990, pp. 1876–80, 1878.

114   'Working Paper for Draft Agreement on Agriculture Trade Proposed by Uruguay Round Agriculture Negotiating Group Chairman Mats Hellstrom', *International Trade Reporter*, 12 December 1990, p. 1905.

115   'U.S., Others Blame EC for Failure in Brussels to Agree on New Rules to Govern World Trade', *International Trade Reporter*, 12 December 1990, pp. 1876–80, 1878; GATT, *GATT Activities 1990*, 1991, p. 26.

116   'Nōsanbutsu, Saitei Yunyūritsu Gimuzuke Teian, Shin Raundo Bunkakaigichō, Hogo Sanwari Sakugen mo' (Uruguay Round Negotiating Group on Agriculture Chair, Agricultural Products, Proposal includes a Minimum Import Percentage Obligation, as well as Thirty Percent Reduction in Protection), *Asahi Shimbun* (evening edition), 7 December 1990, p. 1.

117   'Kome Kaihō, Ketsudan Semarareru Jitai mo, Shin Raundo, Kanbō Chōkan ga Mitōshi' (Uruguay Round, Rice Opening, Chief Cabinet Secretary's Perspective, Possibility that Situation nears for Decision), *Asahi Shimbun*, 8 January 1991, p. 3.

118   'Saitei Yunyū Gimu Ukeire, Tsūshō Daihyō ra Semaru, Gaishō wa Nanshoku' (Foreign Minister Disapproves, USTR and Others Urge Acceptance of Minimum Import Obligation), *Asahi Shimbun*, 16 January 1991, p. 2.

119   'Programme of Work, Proposal by the Chairman at Official Level', GATT, Uruguay Round Trade Negotiations Committee, GATT Doc. No. MTN.TNC/W/69, 26 February 1991, p. 2.

120   'Group of Negotiations on Goods, Nineteenth Meeting: 25 April 1991', GATT, Uruguay Round Group of Negotiations on Goods, GATT Doc. No. MTN.GNG/26, 29 April 1991.

121   'Trade Negotiations Committee, Sixteenth meeting: 25 April 1991', GATT, Uruguay Round Trade Negotiations Committee, GATT Doc. No. MTN.TNC/20, 7 May 1991, pp. 2–3.

122   'Progress of Work in Negotiating Groups: Stock-Taking', GATT, Uruguay Round Trade Negotiations Committee, GATT Doc. No. MTN.TNC/W/89/Add.1, 7 November 1991, pp. 3–4.

123   'Stocktaking of the Uruguay Round Negotiations by the Chairman of the Trade Negotiations Committee at Official Level', GATT, Uruguay Round Trade Negotiations Committee, MTN.TNC/W/89, 7 November 1991, p. 2.

124   'Hills Tells Japanese Leaders to Open Rice Market, Take Lead in GATT Talks', *International Trade Reporter*, 20 November 1991, pp. 1679–80.

125   'Shin Raundo, Zentai no Rieki Kōryo o, Bei Tsūshō Daihyō, Nōsuishō to Kaidan' (Uruguay Round, USTR Meets with Minister of Agriculture, Forestry and Fisheries, Consider All the Advantages), *Asahi Shimbun* (evening edition), 16 November 1991, p. 2.

126   'Trade Negotiations Committee, Twentieth Meeting: 20 December 1991', GATT, Uruguay Round Trade Negotiations Committee, GATT Doc. No. MTN.TNC/24, 9 January 1992, pp. 2–3.

127   'Paragraph 1, Section A: The Calculation of Tariff Equivalents and Related Provisions, Annex 3: Market Access: Agricultural Products subject to

Border Measures Other than Ordinary Customs Duties, Part B: Agreement on Modalities, L. Text on Agriculture', in 'Draft Final Act', GATT, Uruguay Round Trade Negotiations Committee, GATT Doc. No. MTN.TNC/W/FA, 20 December 1991, p. L. 25.

128   'Paragraph 5, Part B, Agreement on Modalities for the Establishment of Specific Binding Commitments under the Reform Programme, L. Text on Agriculture', in 'Draft Final Act', GATT, Uruguay Round Trade Negotiations Committee, GATT Doc. No. MTN.TNC/W/FA, 20 December 1991, p. L. 19.

129   'Raundo Seikō Doryoku de Itti, Kankei Kakuryō Kondankai' (Round Successful, Effort to Agree, Concerned Cabinet Members' Informal Gathering), *Asahi Shimbun* (evening edition), 25 December 1991, p. 1.

130   GATT, *GATT Activities 1992*, 1993, p. 15. See also 'Trade Negotiations Committee, Meeting on 13 January 1992, Opening Statement and Concluding Remarks by the Chairman', GATT, Uruguay Round Trade Negotiations Committee, GATT Doc. No. MTN.TNC/W/99, 15 January 1992.

131   'Trade Negotiations Committee, Twenty-first Meeting: 13 January 1992', GATT Uruguay Round Trade Negotiations Committee, GATT Doc. No. MTN.TNC/25, 5 February 1992, pp. 7–9.

132   'Trade Negotiations Committee, Meeting on 13 January 1992, Opening Statement and Concluding Remarks by the Chairman', GATT, Uruguay Round Trade Negotiations Committee, GATT Doc. No. MTN.TNC/W/99, 15 January 1992, pp. 3–5. See also 'Meeting of 12 March 1992, Note by the Secretariat', GATT, Uruguay Round Group of Negotiations on Goods, GATT Doc. No. MTN.GNG/MA/8, 3 April 1992, p. 1.

133   '"Kome" Ketsudan Sangatsu ni, Shushō Mitōshi, Kigen, Kakkoku mo Okure' (Prime Minister Forecasts 'Rice' Decision in March, Deadline, but Some Countries Are also Late), *Asahi Shimbun* (evening edition), 25 January 1992, p. 1.

134   'EC, Hogo Sakugen Furezu, Shin Raundo, Nōgyō Bunya de Kokubetsuhyō, Nihon, Kome nado Kūhaku' (Uruguay Round, Concerning Agriculture, Each Country's Schedule, EC Doesn't Touch Reductions in Protection, Japan Leaves Out Rice and Other Products), *Asahi Shimbun* (evening edition), 5 March 1992, p. 1.

135   'President Bush Says He Is More Optimistic About Uruguay Round of GATT Trade Talks', *International Trade Reporter*, 15 April 1992, pp. 669–70.

136   'Trade Negotiations Committee, Twenty-Second Meeting: 10 November 1992', GATT, Uruguay Round Trade Negotiations Committee, GATT Doc. No. MTN.TNC/26, 20 November 1992, p. 6.

137   Ibid., p. 3.

138   'Trade Negotiations Committee, Twenty-Fourth Meeting: 18 December 1992', GATT, Uruguay Round Trade Negotiations Committee, GATT Doc. No. MTN.TNC/28, 13 January 1993, pp. 5, 6.

139   'Trade Negotiations Committee, Twenty-Fifth Meeting: 19 January 1993', GATT, Uruguay Round Trade Negotiations Committee, GATT Doc. No. MTN.TNC/29, 3 February 1993, p. 22.

140   'Report on the Uruguay Round', GATT, Uruguay Round Trade Negotiations Committee, GATT Doc. No. MTN.TNC/W/113, 13 July 1993.

141 'G-7 Leaders Set Out Agenda for Growth, Trade, Russian Aid,' *International Trade Reporter*, 14 July 1993, pp. 1174–5, 1174.

142 'Report on the Uruguay Round, Issued July 7, in Tokyo by Trade Ministers of U.S., EC, Canada, and Japan', *International Trade Reporter*, 14 July 1993, pp. 1184–5, 1184.

143 'Trade Negotiations Committee, Twenty-Eighth Meeting: 28 July 1993', GATT, Uruguay Round Trade Negotiations Committee, GATT Doc. No. MTN.TNC/32, 18 August 1993, pp. 1–2, 8–9.

144 'Address by Mr. Peter Sutherland, Director-General, GATT to the Irish Co-Operative Organisation Society's Conference on the International Factors Affecting the Food Sector, Dublin, 11 September 1993', GATT, *News of the Uruguay Round of Multilateral Trade Negotiations*, NUR 065, 14 September 1993, p. 8.

145 'Government Announces It Will Import 200,000 Tons of Rice this Year', *International Trade Reporter*, 6 October 1993, p. 1690.

146 Kyodo, 'Japan Must Open Rice Market: Espy', *Mainichi Daily News*, 9 October 1993, p. 9.

147 'Kome Yunyū, Keizoku Semaru, Nōsuishō ni Bei Nōmuchōkan' (US Agriculture Secretary Urges Minister of Agriculture, Forestry and Fisheries to Continue Rice Imports), *Asahi Shimbun* (evening edition), 12 October 1993, p. 2.

148 'Kome Kanzeika Ukeire Dashin, Seifu, Jisshi, Rokunen Hodo Yūyo, Mazu 4.5% Teido Yunyū ka' (Government Explores Accepting Rice Tariffication, if Put into Effect, About 6 Years Postponement, at First about 4.5% Imported?), *Asahi Shimbun* (evening edition), 15 October 1993, p. 1.

149 'Kome Shijō Kaihō de Nichibei Gōi, Kanzeika o Rokunenkan Yūyo, Saitei Yunyūryō wa Yōnin e' (US–Japan Agreement Concerning Rice Market Opening, Tariffication Postponed for Six Years, Towards Accepting Minimum Import Amount), *Mainichi Shimbun* (evening edition), 14 October 1993, p. 1.

150 ' "Kome Igai" Sokuji Ukeire, Seifu Hōshin, Kome wa Yūyo Saguru' (Government Plan, 'Except for Rice' Accept Immediately, Explore Postponement for Rice), *Nihon Keizai Shimbun*, 19 October 1993, p. 5.

151 'Espy Says Japan Unlikely to Exclude Rice from Uruguay Round Negotiations', *International Trade Reporter*, 3 November 1993, p. 1846.

152 'Trade Negotiations Committee: Thirty-Third Meeting: 19 November 1993', GATT, Uruguay Round Trade Negotiations Committee, GATT Doc. No. MTN.TNC/37, 29 November 1993, pp. 1, 3.

153 'Keiki Shigekisaku ni Zenryoku, Kaiken de Shushō Kyōchō' (Put All Efforts into Business Stimulation Policy, Prime Minister Emphasizes in Press Conference), *Asahi Shimbun*, 22 November 1993, p. 2.

154 'Trade Negotiations Committee, Thirty-Fourth Meeting: 26 November 1993', GATT, Uruguay Round Trade Negotiations Committee, GATT Doc. No. MTN.TNC/38, 2 December 1993, p. 4.

155 'Hosokawa says Formal Announcement on Rice Will be Made this Week', *International Trade Reporter*, 8 December 1993, p. 2055; see also 'Shushō, Jijitsujō Ukeire, Kome Bubun Kaihō no Chōseian' (Prime Minister Receives the Facts, Partial Rice Opening Compromise Plan), *Asahi Shimbun* (evening edition), 7 December 1993, p. 1.

156 'Progress Made in U.S.-EC Trade Negotiations as of Dec. 7', *International Trade Reporter*, 8 December 1993, p. 2042.

157 'Shushō, Jijitsujō Ukeire, Kome Bubun Kaihō no Chōseian' (Prime Minister Receives the Facts, Partial Rice Opening Compromise Plan), *Asahi Shimbun* (evening edition), 7 December 1993, p. 1.

158 'Trade Negotiations Committee, Thirty-Sixth Meeting: 15 December 1993', GATT, Uruguay Round Trade Negotiations Committee, GATT Doc. No. MTN.TNC/40, 21 December 1993, p. 3.

159 'U.S. Backing Puts New Trade Era on Track, But Questions Still Remain', *International Trade Reporter*, 7 December 1994, p. 1894.

6 Conclusion

1 For an earlier attempt to use Putnam's two-level game approach to analyze the Uruguay Round agriculture negotiations, see Avery, 1993a. See also Henry Nau's efforts to examine 'the opportunities and constraints for export interests' in the Uruguay Round (Nau, 1989, p. 15), and Hampson's assertion that, at the international level, during the negotiations, 'insufficient attention was paid to how to make the ambitious package' 'a politically marketable package' at the domestic level (Hampson, 1995, p. 251).

2 'Prepared Statement of Senator Max Baucus', US Congress, Senate Committee on Finance, Subcommittee on International Trade, 3 November 1989, p. 33.

3 Naka seeks to develop 'a more systematic method to describe, explain, and possibly "predict" U.S.–Japanese bilateral trade negotiation processes and their outcomes'. Naka uses 'four systematically layered conceptual models with different levels of analysis emphasizing transgovernmental processes', and 'tries to achieve a more systematic understanding of the political processes' of the Structural Impediments Initiative (Naka, 1996, pp. xiii–xiv).

Select Bibliography

Aaronson, Susan Ariel (1996) *Trade and the American Dream: A Social History of Postwar Trade Policy* (Lexington, Ky.: University Press of Kentucky).

Adams, L. Jerold (1974) *Theory, Law and Policy of Contemporary Japanese Treaties* (Dobbs Ferry, NY: Oceana Publications).

Adams, L. Jerold (1998) 'United States–Japan Trade Relations, The Function of Treaties and Agreements', *Hōgaku Shinpō* (*The Chuo Law Review*), vol. CIV (August), pp. 47–74.

'Agriculture', in Japan Institute of International Affairs, 1989, pp. 121–6.

Aho, C. Michael and Jonathan David Aronson (1985) *Trade Talks: America Better Listen!* (New York: Council on Foreign Relations).

Akaha, Tsuneo (1985) *Japan in Global Ocean Politics* (Honolulu, HI: University of Hawaii Press).

Akaneya, Tatsuo (1992) *Nihon no Gatto Kanyū Mondai, 'Rejīmu Riron' no Bunseki Shikaku ni yoru Jirei Kenkyū* (*The Problem of Japanese Accession to the GATT, A Case Study in Regime Theory*) (Tokyo: University of Tokyo Press).

Akao, Nobutoshi (ed.) (1983) *Japan's Economic Security* (New York: St. Martin's Press).

Allinson, Gary (1989) 'Politics in Contemporary Japan: Pluralist Scholarship in the Conservative Era – A Review Article', *The Journal of Asian Studies*, vol. 48 (May), pp. 324–32.

Allinson, Gary D. and Yasunori Sone (eds) (1993) *Political Dynamics in Contemporary Japan* (Ithaca, NY: Cornell University Press).

Allison, Graham T. and Philip Zelikow (1999) *The Essence of Decision: Explaining the Cuban Missile Crisis*, 2nd edn (New York: Longman).

Anderson, Kym and Yujiro Hayami (1986) *The Political Economy of Agricultural Protection: East Asia in International Perspective* (Winchester, Mass.: Allen & Unwin).

'Appendix B, Survey of Offers' in 'Agriculture' in 'Draft Final Act Embodying the Results of the Uruguay Round of Multilateral Trade Negotiations, Revision', GATT, Uruguay Round Trade Negotiations Committee, GATT Doc. No. MTN. TNC/W/35/Rev. 1, 3 December 1990, pp. 142–54.

Armacost, Michael H. (1996) *Friends or Rivals? The Insider's Account of U.S.–Japan Relations* (New York: Columbia University Press).

Asahi Shimbun (a Japanese-language daily newspaper in Japan), various dates as cited in the Notes and listed in the Index.

Australian Bureau of Agricultural and Resources Economics (1988) *Japanese Agricultural Policies: A Time of Change* (Canberra, Australia: Australian Government Publishing Service Policy Monograph No. 3).

Avery, William P. (ed.) (1993a) *World Agriculture and the GATT: International Political Economy Yearbook, Volume 7* (Boulder, Col.: Lynne Rienner).

Avery, William P. (1993b) 'Agriculture and Free Trade', in Avery (ed.), *World Agriculture and the GATT*, pp. 1–16.

Avery, William P. (1996) 'American Agriculture and Trade Policymaking: Two-Level Bargaining in the NAFTA', *Policy Sciences*, vol. 29, pp. 113–36.

Avery, William P. and David P. Rapkin (1982) *America in a Changing World Political Economy* (New York: Longman).

Averyt, William F., Jr. (1977) *Agropolitics in the European Community: Interest Groups and the Common Agricultural Policy* (New York: Praeger).

Axelrod, Robert (1984) *The Evolution of Cooperation* (New York: Basic Books).

Axelrod, Robert (1997) *The Complexity of Cooperation: Agent-Based Models of Competition and Collaboration* (Princeton, NJ: Princeton University Press).

Balaam, David N. (1986) 'Self-Sufficiency in Japanese Agriculture: Telescoping and Reconciling the Food Security–Efficiency Dilemma', in William P. Browne and Don F. Hadwiger (eds), *World Food Policies: Toward Agricultural Interdependence* (Boulder, Col.: Lynne Rienner), pp. 91–105.

Baldwin, David A. (1985) *Economic Statecraft* (Princeton, NJ: Princeton University Press).

Baldwin, David A. (ed.) (1993) *Neorealism and Neoliberalism: The Contemporary Debate* (New York: Columbia University Press).

Baldwin, Robert E. (1976) *The Political Economy of Postwar U.S. Trade Policy*, New York University Graduate School of Business Administration Center for the Study of Financial Institutions, *The Bulletin*, No. 4.

Baldwin, Robert E. (1985) *The Political Economy of U.S. Import Policy* (Cambridge, Mass.: MIT Press).

Baldwin, Robert E. (1988a) *Trade Policy in a Changing World Economy* (New York: Harvester-Wheatsheaf).

Baldwin, Robert E. (1988b) 'Toward More Efficient Procedures for Multilateral Trade Negotiations', in Robert E. Baldwin, *Trade Policy in a Changing World Economy*, pp. 190–203.

Baldwin, Robert E. and Christopher S. Magee (2000) *Congressional Trade Votes: From NAFTA Approval to Fast-Track Defeat* (Washington, DC: Institute for International Economics).

Barnds, William (ed.) (1979) *Japan and the United States: Challenges and Opportunities* (New York: New York University Press).

Battistini, Lawrence H. (1953) *Japan and America: From Earliest Times to the Present* (Tokyo: Kenkyūsha).

Bayard, Thomas O. and Kimberly Ann Elliott (1994) *Reciprocity and Retaliation in U.S. Trade Policy* (Washington, DC: Institute for International Economics).

Beaumont, Enid F. (1996) 'Domestic Consequences of Internationalization', in Jong S. Jun and Deil S. Wright (eds), *Globalization and Decentralization: Institutional Contexts, Policy Issues, and Intergovernmental Relations in Japan and the United States* (Washington, DC: Georgetown University Press), pp. 374–87.

Bergsten, C. Fred and William R. Cline (1987) *The United States–Japan Economic Problem*, 2nd edn (Washington, DC: Institute for International Economics).

Bergsten, C. Fred and Marcus Noland (1993) *Reconcilable Differences? United States–Japan Economic Conflict* (Washington, DC: Institute for International Economics).

Bergsten, C. Fred, Kimberly Ann Elliott, Jeffrey J. Schott and Wendy E. Takacs (1987) *Auction Quotas and United States Trade Policy* (Washington, DC: Institute for International Economics).

Bernstein, Gail Lee and Haruhiro Fukui (eds) (1988) *Japan and the World: Essays on Japanese History and Politics in Honour of Ishida Takeshi* (New York: St. Martin's Press).

Berton, Peter, Hiroshi Kimura and I. William Zartman (eds) (1999) *International Negotiation: Actors, Structure/Process, Values* (New York: St. Martin's Press).

Bhagwati, Jagdish (1988) *Protectionism* (Cambridge, Mass.: MIT Press).

Bhagwati, Jagdish (1991) *The World Trading System at Risk* (Princeton, NJ: Princeton University Press).

Bhagwati, Jagdish and Hugh T. Patrick (eds) (1990) *Aggressive Unilateralism: America's 301 Trade Policy and the World Trading System* (Ann Arbor, Mich.: University of Michigan Press).

Bhagwati, Jagdish and Robert E. Hudec (eds) (1996) *Fair Trade and Harmonization: Prerequisites for Free Trade?*, 2 vols (Cambridge, Mass.: MIT Press).

Bhagwati, Jagdish and Mathias Hirsch (eds) (1998) *The Uruguay Round and Beyond: Essays in Honor of Arthur Dunkel* (Ann Arbor, Mich.: University of Michigan Press).

Bhala, Raj (1996) *International Trade Law: Cases and Materials* (Charlottesville, Va.: Michie Law Publishers).

Blaker, Michael K. (1977a) 'Probe, Push, and Panic: The Japanese Tactical Style in International Negotiations', in Scalapino (ed.), *The Foreign Policy of Modern Japan*, pp. 55–101.

Blaker, Michael K. (1977b) *Japanese International Negotiating Style* (New York: Columbia University Press).

Blaker, Michael K. (1999) 'Japan Negotiates with the United States on Rice: "No, No, a Thousand Times, No!"', in Berton, Kimura and Zartman (eds), *International Negotiation: Actors, Structure/Process, Values*, pp. 33–61.

Boaz, David (ed.) (1988) *Assessing the Reagan Years* (Washington, DC: Cato Institute).

Bonanno, Alessandro (ed.) (1990) *Agrarian Policies and Agricultural Systems* (Boulder, Col.: Westview Press).

Bovard, James (1989) *The Farm Fiasco* (San Francisco, Calif.: Institute for Contemporary Studies).

Bovard, James (1991) *The Fair Trade Fraud: How Congress Pillages the Consumer and Decimates American Competitiveness* (New York: St. Martin's Press).

Breen, John M. (1993) 'Agriculture', in Terence P. Stewart (ed.), *The GATT Uruguay Round, A Negotiating History (1986–1992) Vol. I*, pp. 125–254.

Breen, John M. (1999) 'Agriculture', in Stewart (ed.), *The GATT Uruguay Round, A Negotiating History (1986–1994), Volume IV, The End Game (Part I)*, pp. 1–179.

Broadbent, Jeffrey (1998) *Environmental Politics in Japan: Networks of Power and Protest* (New York: Cambridge University Press).

Broadbent, Jeffrey and Yoshito Ishio (1998) 'The "Embedded Broker" State: Social Networks and Political Organization in Japan', in W. Mark Fruin (ed.), *Networks, Markets, and the Pacific Rim: Studies in Strategy* (New York: Oxford University Press), pp. 79–108.

Browne, William P. (1988) *Private Interests, Public Policy, and American Agriculture* (Lawrence, Kan.: University Press of Kansas).

Browne, William P. (1995) *Cultivating Congress: Constituents, Issues, and Interests in Agricultural Policymaking* (Lawrence, Kan.: University Press of Kansas).

Browne, William P. and Allan J. Cigler (eds) (1990) *U.S. Agricultural Groups: Institutional Profiles* (New York: Greenwood Press).

Buckley, Roger (1992) *US–Japan Alliance Diplomacy, 1945–1990* (New York: Cambridge University Press).

Bueno de Mesquita, Bruce (2000) *Principles of International Politics: People's Power, Preferences, and Perceptions* (Washington, DC: CQ Press).

Busch, Marc L. (1999) *Trade Warriors: States, Firms, and Strategic Policy in High Technology* (New York: Cambridge University Press).

Bush, George (1990, 1991, 1992, 1993) *Public Papers of the Presidents of the United States: George Bush* (Washington, DC: Government Printing Office).

Caldas, Ricardo Wahrendorff (1998) *Brazil in the Uruguay Round of the GATT: The Evolution of Brazil's Position in the Uruguay Round, with Emphasis on the Issue of Services* (Brookfield, Vt.: Ashgate).

Calder, Kent E. (1988a) 'Japanese Foreign Economic Policy Formation: Explaining the Reactive State', *World Politics*, vol. 40 (July), pp. 517–41.

Calder, Kent E. (1988b) *Crisis and Compensation: Public Policy and Political Stability in Japan: 1949–1986* (Princeton, NJ: Princeton University Press).

Callon, Scott (1995) *Divided Sun: MITI and the Breakdown of Japanese High-Tech Industrial Policy, 1975–1993* (Stanford, Calif.: Stanford University Press).

Caporaso, James A. (1997) 'Across the Great Divide: Integrating Comparative and International Politics', *International Studies Quarterly*, vol. 41 (December), pp. 563–91.

Cardoso, Fernando Henrique (1973) 'Associated Dependent Development: Theoretical and Practical Implications', in Alfred Stepan (ed.), *Authoritarian Brazil* (New Haven, Conn.: Yale University Press), pp. 142–76.

Castle, Emery N. and Kenzo Hemmi, with Sally A. Skillings (eds) (1982) *U.S.–Japanese Agricultural Trade Relations* (Washington, DC: Resources for the Future).

Choate, Pat and Juyne Linger (1988) 'Tailored Trade: Dealing with the World As It Is', *Harvard Business Review*, vol. 66 (January–February), pp. 86–93.

Choy, Jon K. T. (ed.) (1988) *Japan: Exploring New Paths* (Washington, DC: Japan Economic Institute).

Christensen, Ray (2000) *Ending the LDP Hegemony: Party Cooperation in Japan* (Honolulu, HI: University of Hawaii Press).

Cigler, Allan J. (1991) 'Organizational Maintenance and Political Activity on the "Cheap": The American Agriculture Movement', in Allan J. Cigler and Burdett A. Loomis (eds), *Interest Group Politics*, 3rd edn (Washington, DC: CQ Press), pp. 81–107.

Cigler, Allan J. and Burdett A. Loomis (eds) (1995) *Interest Group Politics*, 4th edn (Washington, DC: CQ Press).

Clinton, William J. (1994, 1995) *Public Papers of the Presidents of the United States: William J. Clinton* (Washington, DC: Government Printing Office).

Cohen, Benjamin J. (1990) 'The Political Economy of International Trade', *International Organization*, vol. 44 (Spring), pp. 261–81.

Cohen, Stephen D. (1985) *Uneasy Partnership: Competition and Conflict in U.S.–Japanese Trade Relations* (Cambridge, Mass.: Ballinger Publishing).

Cohen, Stephen D. (1991) *Cowboys and Samurai: Why the United States Is Losing the Industrial Battle and Why It Matters* (New York: Harper Business).

Cohen, Stephen D. (1998) *An Ocean Apart: Explaining Three Decades of U.S.–Japanese Trade Frictions* (Westport, Conn.: Praeger).

Cohen, Stephen D. (2000) *The Making of United States International Economic Policy: Principles, Problems, and Proposals for Reform*, 5th edn (Westport, Conn.: Praeger).

Cohen, Stephen D., Joel R. Paul and Robert A. Blecker (1996) *Fundamentals of U.S. Foreign Trade Policy: Economics, Politics, Laws, and Issues* (Boulder, Col.: Westview Press).

Cohn, Theodore H. (1993a) 'The Changing Role of the United States in the Global Agricultural Trade Regime', in Avery (ed.), *World Agriculture and the GATT*, pp. 17–38.

Cohn, Theodore H. (1993b) 'The Intersection of Domestic and Foreign Policy in the NAFTA Agricultural Negotiations', *Canadian–American Public Policy*, vol. 14 (September).

Collier, David (ed.) (1979) *The New Authoritarianism in Latin America* (Princeton, NJ: Princeton University Press).

'Constitution of Japan (*Nihonkoku Kempō*)' (1991) in Inoue, pp. 271–99.

Conybeare, John A. C. (1987) *Trade Wars: The Theory and Practice of International Commercial Rivalry* (New York: Columbia University Press).

Conybeare, John A. C. (1991) 'Voting for Protection: An Electoral Model of Tariff Policy', *International Organization*, vol. 45 (Winter), pp. 57–81.

Cooper, Richard N. (1972–3) 'Trade Policy is Foreign Policy', *Foreign Policy*, vol. 9 (Winter), pp. 18–36.

Cortell, Andrew P. and James W. Davis, Jr. (1996) 'How Do International Institutions Matter? The Domestic Impact of International Rules and Norms', *International Studies Quarterly*, vol. 40, pp. 451–78.

Cowhey, Peter F. (1993a) 'Domestic Institutions and the Credibility of International Commitments: Japan and the United States', *International Organization*, vol. 47 (Spring), pp. 299–326.

Cowhey, Peter F. (1993b) 'Elect Locally – Order Globally: Domestic Politics and Multilateral Cooperation', in Ruggie (ed.), *Multilateralism Matters*, pp. 157–200.

Cowhey, Peter F. (1995) 'The Politics of Foreign Policy in Japan and the United States', in Cowhey and McCubbins (eds), *Structure and Policy in Japan and the United States*, pp. 203–25.

Cowhey, Peter F. and Jonathan D. Aronson (1993) *Managing the World Economy: The Consequences of Corporate Alliances* (New York: Council on Foreign Relations).

Cowhey, Peter F. and Edward Long (1983) 'Testing Theories of Regime Change: Hegemonic Decline or Surplus Capacity?', *International Organization*, vol. 37 (Spring), pp. 157–88.

Cowhey, Peter F. and Mathew D. McCubbins (eds) (1995) *Structure and Policy in Japan and the United States* (New York: Cambridge University Press).

Coyle, William T., Dermot Hayes and Hiroshi Yamauchi (eds) (1992) *Agriculture and Trade in the Pacific: Toward the Twenty-First Century* (Boulder, Col.: Westview Press).

Crabb, Cecil V., Jr. and Pat M. Holt (1992) *Invitation to Struggle: Congress, the President, and Foreign Policy*, 4th edn (Washington, DC: Congressional Quarterly Press).

Crawcour, E. Sydney (1997) 'Economic Change in the Nineteenth Century', in Kozo Yamamura (ed.), *The Economic Emergence of Modern Japan* (New York: Cambridge University Press), pp. 1–49.

Croome, John (1995) *Reshaping the World Trading System: A History of the Uruguay Round* (Geneva: World Trade Organization).

Curran, T. J. (1982) 'Politics and High Technology: The NTT Case,' in Destler and Sato (eds), *Coping with U.S.–Japanese Economic Conflicts*, pp. 185–241.

Curran, T. J. (1983) 'Politics of Trade Liberalization in Japan', *Journal of International Affairs*, vol. 37 (Summer), pp. 105–22.

Curtis, Gerald L. (ed.) (1970) *Japanese–American Relations in the 1970s* (Washington, DC: Columbia Books).

Curtis, Gerald L. (1975) 'Big Business and Political Influence', in Ezra F. Vogel (ed.), *Modern Japanese Organization and Decision-Making* (Berkeley, Calif.: University of California Press), pp. 33–70.

Curtis, Gerald L. (1979) 'Domestic Politics and Japanese Foreign Policy', in Barnds (ed.), *Japan and the United States*, pp. 21–85.

Curtis, Gerald L. (1988) *The Japanese Way of Politics* (New York: Columbia University Press).

Curtis, Gerald L. (ed.) (1993) *Japan's Foreign Policy After the Cold War: Coping with Change* (Armonk, NY: M. E. Sharpe).

Curtis, Gerald L. (ed.) (1994) *The United States, Japan, and Asia: Challenges for U.S. Policy* (New York: W. W. Norton).

Curtis, Gerald L. (1999) *The Logic of Japanese Politics: Leaders, Institutions, and the Limits of Change* (New York: Columbia University Press).

Curzon, Gerard and Victoria Curzon (1973) 'GATT: Traders' Club', in Robert W. Cox and Harold K. Jacobson and Gerard and Victoria Curzon, Joseph S. Nye, Lawrence Scheinman, James P. Sewell, and Susan Strange, *The Anatomy of Influence: Decision Making in International Organization* (New Haven, Conn.: Yale University Press), pp. 298–333.

Dahl, R. A. (1956) *A Preface to Democratic Theory* (Chicago, Ill.: University of Chicago Press).

Dahl, R. A. (1971) *Polyarchy: Participation and Opposition* (New Haven, Conn.: Yale University Press).

Dahl, R. A. (1985) *A Preface to Economic Democracy* (Berkeley, Calif.: University of California Press).

Dam, Kenneth W. (1970) *The GATT: Law and International Economic Organization* (Chicago, Ill.: University of Chicago Press).

Deardorff, Alan V. and Robert M. Stern (eds) (1994) *Analytical and Negotiating Issues in the Global Trading System* (Ann Arbor, Mich.: University of Michigan Press).

Deardorff, Alan V. and Robert M. Stern (eds) (2000) *Social Dimensions of U.S. Trade Policies* (Ann Arbor, Mich.: University of Michigan Press).

Destler, I. M. (1980) *Making Foreign Economic Policy* (Washington, DC: Brookings Institution).

Destler, I. M. (1994) 'Delegating Trade Policy', in Paul E. Peterson (ed.), *The President, the Congress and the Making of Foreign Policy* (Norman, Okla.: University of Oklahoma Press), pp. 228–45.

Destler, I. M. (1995) *American Trade Politics*, 3rd edn (Washington, DC: Institute for International Economics and New York: Twentieth Century Fund).

Destler, I. M. (1997a) 'Has Conflict Passed Its Prime? Japanese and American Approaches to Trade and Economic Policy', Maryland/Tsukuba Papers on U.S.–Japan Relations (College Park, Md.: Center for International and Security Studies at Maryland, March).

Destler, I. M. (1997b) *Renewing Fast-Track Legislation* (Washington, DC: Institute for International Economics).

Destler, I. M. and Hideo Sato (eds) (1982a) *Coping with U.S.–Japanese Economic Conflicts* (Lexington, Mass.: Lexington Books).

Destler, I. M. and Hideo Sato (1982b) 'Coping with Economic Conflicts', in Destler and Sato (eds), *Coping with U.S.–Japanese Economic Conflicts*, pp. 271–93.

Destler, I. M. and Michael Nacht (1992) 'Beyond Mutual Recrimination: Building a Solid U.S.–Japan Relationship in the 1990s', in Sean M. Lynn-Jones and Steven E. Miller (eds), *America's Strategy in a Changing World* (Cambridge, Mass.: MIT Press), pp. 267–94.

Destler, I. M. and Peter J. Balint (1999) *The New Politics of American Trade: Trade, Labor, and the Environment* (Washington, DC: Institute for International Economics).

Destler, I. M., Hideo Sato, Priscilla Clapp and Haruhiro Fukui (1976) *Managing an Alliance: The Politics of U.S.–Japanese Relations* (Washington, DC: Brookings Institution).

Destler, I. M., Haruhiro Fukui and Hideo Sato (1979) *The Textile Wrangle: Conflict in Japanese–American Relations, 1969–1971* (Ithaca, NY: Cornell University Press).

Destler, I. M. and John S. Odell, assisted by Kimberly Ann Elliott (1987) *Anti-Protection: Changing Forces in United States Trade Politics* (Washington, DC: Institute for International Economics).

Donnelly, Michael Wade (1978) 'Political Management of Japan's Rice Economy', Ph.D. dissertation, Columbia University.

Donnelly, Michael Wade (1984) 'Conflict over Government Authority and Markets: Japan's Rice Economy', in Ellis S. Krauss, Thomas P. Rohlen and Patricia G. Steinhoff (eds), *Conflict in Japan* (Honolulu, HI: University of Hawaii Press), pp. 335–74.

Donnelly, Michael Wade (1993) 'On Political Negotiation: America Pushes to Open Up Japan', *Pacific Affairs*, vol. 66 (Fall), pp. 329–50.

Donnelly, Michael Wade (1998) 'Japanese–American Trade Negotiations: The Structural Impediments Initiative', in Michael Fry, John Kirton and Mitsuru Kurosawa (eds), *The North Pacific Triangle: The United States, Japan, and Canada at Century's End* (Toronto: University of Toronto Press), pp. 60–84.

Downs, George W. and David M. Rocke (1995) *Optimal Imperfection? Domestic Uncertainty and Institutions in International Relations* (Princeton, NJ: Princeton University Press).

Doyle, Michael W. and G. John Ikenberry (eds) (1997) *New Thinking in International Relations Theory* (Boulder, Col.: Westview Press).

Drake, William J. and Kalypso Nicolaidis (1992) 'Ideas, Interests, and Institutionalization: "Trade in Services" and the Uruguay Round', *International Organization*, vol. 46 (Winter), pp. 37–100.

Drifte, Reinhard (1998) *Japan's Foreign Policy for the 21st Century: From Economic Superpower to What Power?*, 2nd edn (New York: St. Martin's Press).

Dryden, Steve (1995) *Trade Warriors: USTR and the American Crusade for Free Trade* (New York: Oxford University Press).

Dunkel, Arthur (1990) 'Introduction by the Director-General of GATT', in GATT, *GATT Activities 1989*, pp. 7–10.

Eckes, Alfred E., Jr. (1995) *Opening America's Market: U.S. Foreign Trade Policy since 1776* (Chapel Hill, NC: University of North Carolina Press).

Eckes, Alfred E., Jr. (1999) 'U.S. Trade History', in Lovett, Eckes and Brinkman (eds), *U.S. Trade Policy*, pp. 51–105.

Epstein, David and Sharyn O'Halloran (1996) 'The Partisan Paradox and the U.S. Tariff, 1877–1934', *International Organization*, vol. 50 (Spring), pp. 301–24.

Epstein, David and Sharyn O'Halloran (1999) *Delegating Powers: A Transaction Cost Politics Approach to Policy Making under Separate Powers* (New York: Cambridge University Press).

Evangelista, Matthew (1997) 'Domestic Structure and International Change', in Michael W. Doyle and G. John Ikenberry (eds), *New Thinking in International Relations Theory* (Boulder, Col.: Westview Press), pp. 202–28.

Evans, Peter B. (1979) *Dependent Development: The Alliance of Multinational, State, and Local Capital in Brazil* (Princeton, NJ: Princeton University Press).

Evans, Peter B. (1993) 'Building an Integrative Approach to International and Domestic Politics: Reflections and Projections', in Evans, Jacobson and Putnam (eds), *Double-Edged Diplomacy*, pp. 397–430.

Evans, Peter B. (1995) *Embedded Autonomy: States and Industrial Transformation* (Princeton, NJ: Princeton University Press).

Evans, Peter B., Dietrich Rueschemeyer and Theda Skocpol (eds) (1985) *Bringing the State Back In* (New York: Cambridge University Press).

Evans, Peter B., Harold K. Jacobson and Robert D. Putnam (eds) (1993) *Double-Edged Diplomacy: International Bargaining and Domestic Politics* (Berkeley, Calif.: University of California Press).

Faini, Riccardo and Enzo Grilli (eds) (1997) *Multilateralism and Regionalism after the Uruguay Round* (New York: St. Martin's Press).

Fairbank, John K., Edwin O. Reischauer and Albert M. Craig (1989) *East Asia: Tradition and Transformation*, revised edn (Boston, Mass.: Houghton Mifflin).

Fearon, James (1998) 'Bargaining, Enforcement and International Cooperation', *International Organization*, vol. 52 (Spring), pp. 269–305.

Federal Reserve Bank of Kansas City (1985) *Competing in the World Marketplace: The Challenge for American Agriculture: A Symposium* (Kansas City, Mo.: Federal Reserve Bank of Kansas City).

Feldstein, Martin (ed.) (1988a) *International Economic Cooperation* (Chicago, Ill.: University of Chicago Press).

Feldstein, Martin (ed.) (1988b) *The United States in the World Economy* (Chicago, Ill.: University of Chicago Press).

Feldstein, Martin (ed.) (1994) *American Economic Policy in the 1980s* (Chicago, Ill.: University of Chicago Press).

Feller, Peter Buck and Ann Carlisle Wilson (1996) 'United States Tariff and Trade Law: Constitutional Sources and Constraints', *Law and Policy in International Business*, vol. 8, pp. 105–23.

Finlayson, Jock A. and Mark W. Zacher (1982) 'The GATT and the Regulation of Trade Barriers: Regime Dynamics and Functions', in Krasner (ed.), *International Regimes*, pp. 273–314.

'First "Maekawa" Report, The Report of the Advisory Group on Economic Structural Adjustment for International Harmony, Submitted to the Prime Minister April 7, 1986' (1988) trans. the Embassy of Japan, Washington, DC, in Choy, pp. 9–13.

Flamm, Kenneth (1996) *Mismanaged Trade? Strategic Policy and the Semiconductor Industry* (Washington, DC: Brookings Institution).

Fletcher, William Miles III (1989) *The Japanese Business Community and National Trade Policy, 1920–1942* (Chapel Hill, NC: University of North Carolina Press).

Food and Agriculture Policy Research Center (FAPRC) Study Group on International Issues (SGII) (1990a) 'Japanese Beef Industry Facing Trade Liberalization' (Tokyo: FAPRC Japan SGII Report, no. 6, June).

Food and Agriculture Policy Research Center (FAPRC) Study Group on International Issues (SGII) (1990b) 'Agricultural Policies of Japan', (Tokyo: FAPRC Japan SGII Report, no. 9, December).

Forum for Policy Innovation (Seisaku Kōsō Forum) (1990) 'Kanzeika ni yoru Kome no Shijō Kaihō o (Toward Tariffication for opening the Rice Market in Japan)' (Tokyo: Forum for Policy Innovation, July).

Frank, Andre Gunder (1967) *Capitalism and Underdevelopment in Latin America* (New York: Monthly Review Press).

Frank, Isaiah (ed.) (1975) *The Japanese Economy in International Perspective* (Baltimore, Md.: Johns Hopkins University Press).

Frank, Isaiah and Ryokichi Hirono (eds) (1974) *How the United States and Japan See Each Other's Economy* (New York: Committee for Economic Development).

Fried, Edward R., Jimmye Hillman, D. Gale Johnson, Philip H. Trezise, T. K. Warley, Francesco De Stefano, Wolfgang Hager, B. Heringa, Max Kohnstamm, John Pinder, Adrien Zeller, Yujiro Hayami, Kenzo Hemmi, and Hisao Kanamori (1973) *Toward the Integration of World Agriculture: A Tripartite Report by Fourteen Experts from North America, the European Community, and Japan* (Washington, DC: Brookings Institution).

Frieden, Jeff (1988) 'Sectoral Conflict and Foreign Economic Policy, 1914–1940', *International Organization*, vol. 42 (Winter), pp. 59–90.

Friman, H. Richard (1990) *Patchwork Protectionism: Textile Trade Policy in the United States, Japan, and West Germany* (Ithaca, NY: Cornell University Press).

Friman, H. Richard (1993) 'Side-Payments versus Security Cards: Domestic Bargaining Tactics in International Economic Negotiations', *International Organization*, vol. 47 (Summer), pp. 387–410.

Fruin, W. Mark (1992) *The Japanese Enterprise System: Competitive Strategies and Cooperative Structures* (New York: Oxford University Press).

Fukui, Haruhiro (1977) 'Studies in Policymaking: A Review of the Literature', in Pempel (ed.), *Policymaking in Contemporary Japan*, pp. 22–59.

Fukui, Haruhiro (1987) 'The Policy Research Council of Japan's Liberal Democratic Party: Policy Making Role and Practice', *Asian Thought and Society*, vol. 12 (March), pp. 3–31.

Gardner, Bruce L. (1986) 'Farm Policy and the Farm Problem', in Philip Cagan (ed.), *Essays in Contemporary Economic Problems, 1986: The Impact of the Reagan Program* (Washington, DC: American Enterprise Institute), pp. 223–46.

Gardner, Richard N. (1980) *Sterling–Dollar Diplomacy in Current Perspective: The Origins and the Prospects of our International Economic Order*, new edn (New York: Columbia University Press).

Garten, Jeffrey E. (1992) *A Cold Peace: America, Japan, Germany, and the Struggle for Supremacy* (New York: Times Books).

GATT (1983–96) *GATT Activities, An Annual Review of the Work of the GATT* (Geneva: General Agreement on Tariffs and Trade).

GATT (1985) *Trade Policies for a Better Future: Proposals for Action* (Geneva: General Agreement on Tariffs and Trade).

GATT (1990a) *Trade Policy Review Series, United States, 1989* (Geneva: GATT Publication Services).

GATT (1990b) *Trade Policy Review Series, Japan, 1990* (Geneva: GATT Publication Services).

GATT (1993) *Trade Policy Review Series, Japan, 1992* (Geneva: GATT Publication Services).

GATT (1994a) *Uruguay Round of Multilateral Trade Negotiations, Legal Instruments Embodying the Results of the Uruguay Round of Multilateral Trade Negotiations Done at Marrakesh on 15 April 1994* (Geneva: GATT Secretariat).

GATT (1994b) *The Legal Texts: The Results of the Uruguay Round of Multilateral Trade Negotiations* (Geneva: GATT Secretariat).

GATT (1994c) *Analytical Index: Guide to GATT Law and Practice*, 6th edn (Geneva: General Agreement on Tariffs and Trade).

GATT (1995) *GATT Trade Policy Review Series, Japan, 1994* (Geneva: GATT Publication Services).

GATT (1997–2000) *Uruguay Round Consolidated List and Index of Documents Issued between 1986 and 1994* (prepared for microfiche users only) (Geneva: General Agreement on Tariffs and Trade).

GATT, *News of the Uruguay Round of Multilateral Trade Negotiations* (1988a) 'Agriculture . . . 9 and 10 June', NUR 017, 30 June, pp. 6–7.

GATT, *News of the Uruguay Round of Multilateral Trade Negotiations* (1988b) 'Agriculture . . . 13 and 14 July', NUR 018, 2 August, pp. 9–11.

GATT, *News of the Uruguay Round of Multilateral Trade Negotiations* (1988c) 'Agriculture . . . 12 and 13 September', NUR 019, 5 October, pp. 2/3–4.

GATT, *News of the Uruguay Round of Multilateral Trade Negotiations* (1988d) 'Trade in Agricultural Products . . . 11–14 October', NUR 020, 4 November, pp. 6–8.

GATT, *News of the Uruguay Round of Multilateral Trade Negotiations* (1989a) 'Agriculture . . . 10 to 12 July', NUR 030, 3 August, pp. 7–9.

GATT, *News of the Uruguay Round of Multilateral Trade Negotiations* (1989b) 'Agriculture . . . 25–26 September', NUR 031, 16 October, pp. 8–9.

GATT, *News of the Uruguay Round of Multilateral Trade Negotiations* (1989c) 'Agriculture . . . 25–26 October', NUR 032, 21 November, pp. 6–7.

GATT, *News of the Uruguay Round of Multilateral Trade Negotiations* (1990a) 'Collective Determination to Succeed but Negotiations behind Schedule, Notes Dunkel at TNC', NUR 039, 30 July.

GATT, *News of the Uruguay Round of Multilateral Trade Negotiations* (1990b) 'Agriculture . . . 27–29 August', NUR 041, 9 October, pp. 1–4.

GATT, *News of the Uruguay Round of Multilateral Trade Negotiations* (1993a) 'Address by Mr. Peter Sutherland, Director-General, GATT to the Irish Co-operative Organisation Society's Conference on the International Factors Affecting the Food Sector, Dublin, 11 September 1993', NUR 065, 14 September, pp. 4–10.

GATT, *News of the Uruguay Round of Multilateral Trade Negotiations* (1993b) 'The Final Act of the Uruguay Round: Press Summary', NUR 080, 14 December 1993.

GATT, Uruguay Round Group of Negotiations on Goods (1987a) 'Fifth Meeting of the Group of Negotiations on Goods: Record of Decisions Taken', GATT Doc. No. MTN.GNG/5, 9 February.

GATT, Uruguay Round Group of Negotiations on Goods (1987b) 'Chairmen of Negotiating Groups', GATT Doc. No. MTN.GNG/W/9, 12 February.

GATT, Uruguay Round Group of Negotiations on Goods (1988a) 'Group of Negotiations on Goods, Eleventh meeting: 25 and 26 July 1988', GATT Doc. No. MTN.GNG/12, 15 August.

GATT, Uruguay Round Group of Negotiations on Goods (1988b) 'Report by the Chairman of the Negotiating Group on Agriculture, Part B – Points for Decision', GATT Doc. No. MTN.GNG/16, 30 November.

GATT, Uruguay Round Group of Negotiations on Goods (1990) 'Group of Negotiations on Goods, Sixteenth meeting: 9 April 1990', GATT Doc. No. MTN.GNG/22, 8 May.

GATT, Uruguay Round Group of Negotiations on Goods (1991) 'Group of Negotiations on Goods, Nineteenth meeting: 25 April 1991', GATT Doc. No. MTN.GNG/26, 29 April.

GATT, Uruguay Round Group of Negotiations on Goods (1992) 'Meeting of 12 March 1992, Note by the Secretariat', GATT Doc. No. MTN.GNG/MA/8, 3 April.

GATT, Uruguay Round Negotiating Group on Agriculture (1987a) 'Summary of Major Problems and their Causes as Identified Thus Far and of Issues Considered Relevant, Note by the Secretariat', GATT Doc. No. MTN.GNG/NG5/W/2, 31 March.

GATT, Uruguay Round Negotiating Group on Agriculture (1987b) 'Summary of the Main Points Raised in the Course of the Group's Consideration of Basic Principles to Govern World Trade in Agriculture: 5–6 May 1987, Note by the Secretariat', GATT Doc. No. MTN.GNG/NG5/W/12, 22 June.

GATT, Uruguay Round Negotiating Group on Agriculture (1987c) 'Summary of Studies on Problems Affecting Trade in Agriculture and their Causes, Note by the Secretariat, Addendum, OECD: National Policies and Agricultural Trade', GATT Doc. No. MTN.GNG/NG5/W/3/Add.1, 25 June.

GATT, Uruguay Round Negotiating Group on Agriculture (1987d) 'List of Representatives', GATT Doc. No. MTN.GNG/NG5/INF/1, 26 October.

GATT, Uruguay Round Negotiating Group on Agriculture (1987e) 'Summary of Main Points Raised at the Fourth Meeting of the Negotiating Group on Agriculture (26–27 October 1987), Note by the Secretariat', GATT Doc. No. MTN.GNG/NG5/W/33, 19 November.

GATT, Uruguay Round Negotiating Group on Agriculture (1988a) 'Summary of Main Points Raised at the Sixth Meeting of the Negotiating Group on Agriculture (15–17 February 1988), Note by the Secretariat', GATT Doc. No. MTN.GNG/NG5/W/52, 17 March.

GATT, Uruguay Round Negotiating Group on Agriculture (1988b) 'Summary of Main Points Raised at the Eighth Meeting of the Negotiating Group on Agriculture (9–10 June 1988), Note by the Secretariat', GATT, Uruguay Round Negotiating Group on Agriculture, GATT Doc. No. MTN.GNG/NG5/W/63, 28 June.

GATT, Uruguay Round Negotiating Group on Agriculture (1988c) 'Summary of Main Points Raised at the Ninth Meeting of the Negotiating Group on Agriculture (13–14 July 1988), Note by the Secretariat', GATT Doc. No. MTN.GNG/NG5/W/73, 29 July.

GATT, Uruguay Round Negotiating Group on Agriculture (1988d) 'Summary of Main Points Raised at the Tenth Meeting of the Negotiating Group on Agriculture (12–13 September 1988), Note by the Secretariat', GATT Doc. No. MTN.GNG/NG5/W/79, 29 September.

GATT, Uruguay Round Negotiating Group on Agriculture (1988e) 'Summary of Main Points Raised at the Eleventh Meeting of the Negotiating Group on Agriculture (13–14 October 1988), Note by the Secretariat', GATT Doc. No. MTN.GNG/NG5/W/86, 10 November.

GATT, Uruguay Round Negotiating Group on Agriculture (1989a) 'GATT Rules and Disciplines Relating to Agriculture, Note by the Secretariat', GATT Doc. No. MTN.GNG/NG5/W/95, 4 July.

GATT, Uruguay Round Negotiating Group on Agriculture (1989b) 'Summary of the Main Points Raised at the Fourteenth Meeting of the Negotiating Group on Agriculture (10–12 July 1989), Note by the Secretariat', GATT Doc. No. MTN.GNG/NG5/W/103, 4 September.

GATT, Uruguay Round Negotiating Group on Agriculture (1989c) 'Summary of the Main Points Raised at the Fifteenth Meeting of the Negotiating Group on Agriculture (25–26 September 1989) Note by the Secretariat', GATT Doc. No. MTN.GNG/NG5/W/113, 23 October.

GATT, Uruguay Round Negotiating Group on Agriculture (1989d) 'Summary of the Main Points Raised at the Sixteenth Meeting of the Negotiating Group on Agriculture (25–26 October 1989) Note by the Secretariat', GATT Doc. No. MTN.GNG/NG5/W/123, 10 November.

GATT, Uruguay Round Negotiating Group on Agriculture (1989e) 'Summary of the Main Points Raised at the Seventeenth Meeting of the Negotiating Group on Agriculture (27–28 November 1989) Note by the Secretariat', GATT Doc. No. MTN.GNG/NG5/W/139, 14 December.

GATT, Uruguay Round Negotiating Group on Agriculture (1990a) 'Synoptic Table of Negotiating Proposals Submitted pursuant to Paragraph 11 of the Mid-Term Review Agreement on Agriculture, Note by the Secretariat', GATT Doc. No. MTN.GNG/NG5/W/150, 12 February.

GATT, Uruguay Round Negotiating Group on Agriculture (1990b) 'Summary of the Main Points Raised at the Eighteenth Meeting of the Negotiating Group on Agriculture (19–20 December 1989) Note by the Secretariat', GATT Doc. No. MTN.GNG/NG5/W/149, 6 February.

GATT, Uruguay Round Negotiating Group on Agriculture (1990c) 'Synoptic Table of Negotiating Proposals Submitted pursuant to Paragraph 11 of the Mid-Term Review Agreement on Agriculture, Note by the Secretariat, Revision', GATT Doc. No. MTN.GNG/NG5/W/150/Rev.1, 2 April.

GATT, Uruguay Round Negotiating Group on Agriculture (1990d) 'Clarification and Elaboration of Elements of Detailed Proposals Submitted Pursuant to the Mid-Term Review Decision, Note by the Secretariat', GATT Doc. No. MTN.GNG/NG5/W/161, 4 April.

GATT, Uruguay Round Negotiating Group on Agriculture (1990e) 'Clarification and Elaboration of Elements of Detailed Proposals Submitted Pursuant to the

Mid-Term Review Decision, Note by the Secretariat, Addendum', GATT Doc. No. MTN.GNG/NG5/W/161/Add.2, 18 May.

GATT, Uruguay Round Negotiating Group on Agriculture (1990f) 'Framework Agreement on Agriculture Reform Programme, Draft Text by the Chairman', GATT Doc. No. MTN.GNG/NG5/W/170, 11 July.

GATT, Uruguay Round Negotiating Group on Agriculture (1990g) 'Santiago Meeting of Cairns Group Ministers: 4–6 July 1990, Press Communique and Conclusions, Submitted by Australia on Behalf of the Cairns Group', GATT Doc. No. MTN.GNG/NG5/W/175, 24 July.

GATT, Uruguay Round Negotiating Group on Agriculture (1990h) 'Twenty-Fifth Session of the Negotiating Group on Agriculture, Note by the Chairman', GATT Doc. No. MTN.GNG/NG5/26, 4 October.

GATT, Uruguay Round Negotiating Group on Agriculture (1990i) 'Twenty-Fifth (Resumed) Session of the Negotiating Group on Agriculture, Note by the Chairman, Revision', GATT Doc. No. MTN.GNG/NG5/27/Rev.1, 29 October.

GATT, Uruguay Round Trade Negotiations Committee (1986) 'List of Representatives – Group of Negotiations on Goods – Group of Negotiations on Services', GATT Doc. No. MTN.TNC/INF/1, MTN.GNG/INF1/1, 27 October.

GATT, Uruguay Round Trade Negotiations Committee (1987) 'Annex: Attendance of International Organizations in the Proceedings of the Uruguay Round, Adopted by the Trade Negotiations Committee, 3 July 1987, Trade Negotiations Committee, Third Meeting: 3 July 1987', GATT Doc. No. MTN.TNC/3, 22 July.

GATT, Uruguay Round Trade Negotiations Committee (1988a) 'Meeting at Ministerial Level, December 1988: Checklist of documents, Revision', GATT Doc. No. MTN.TNC/W/13/Rev.1, 1 December.

GATT, Uruguay Round Trade Negotiations Committee (1988b) 'Trade Negotiations Committee Meeting at Ministerial Level, Montreal, December 1988', GATT Doc. No. MTN.TNC/7(MIN), 9 December.

GATT, Uruguay Round Trade Negotiations Committee (1989a) 'Meeting at Ministerial Level, Palais des Congrès, Montreal (Canada), 5–9 December 1988', GATT Doc. No. MTN.TNC/8(MIN), 17 January.

GATT, Uruguay Round Trade Negotiations Committee (1989b) 'List of Representatives', GATT Doc. No. MTN.TNC/INF/6/Rev.1, 6 April.

GATT, Uruguay Round Trade Negotiations Committee (1989c) 'Trade Negotiations Committee Meeting at Level of High Officials, Geneva, 5–8 April 1989', GATT Doc. No. MTN.TNC/9, 11 April.

GATT, Uruguay Round Trade Negotiations Committee (1989d) 'Trade Negotiations Committee, Ninth Meeting: 27 July 1989', GATT Doc. No. MTN.TNC/12, 16 August.

GATT, Uruguay Round Trade Negotiations Committee (1990a) 'Trade Negotiations Committee, Chairman's Summing-up at meeting of 26 July 1990', GATT Doc. No. MTN.TNC/15, 30 July.

GATT, Uruguay Round Trade Negotiations Committee (1990b) 'Draft Final Act Embodying the Results of the Uruguay Round of Multilateral Trade Negotiations, Revision', GATT Doc. No. MTN.TNC/W/35/Rev.1, 3 December.

GATT, Uruguay Round Trade Negotiations Committee (1991a) 'Programme of Work, Proposal by the Chairman at Official Level', GATT Doc. No. MTN.TNC/W/69, 26 February.

GATT, Uruguay Round Trade Negotiations Committee (1991b) 'Trade Negotiations Committee, Sixteenth Meeting: 25 April 1991', GATT Doc. No. MTN.TNC/20, 7 May.

GATT, Uruguay Round Trade Negotiations Committee (1991c) 'Stock-Taking of the Uruguay Round Negotiations by the Chairman of the Trade Negotiations Committee at Official Level', GATT Doc. No. MTN.TNC/W/89, 7 November.

GATT, Uruguay Round Trade Negotiations Committee (1991d) 'Progress of Work in Negotiating Groups: Stock-Taking', GATT Doc. No. MTN.TNC/W/89/Add.1, 7 November.

GATT, Uruguay Round Trade Negotiations Committee (1991e) 'Draft Final Act Embodying the Results of the Uruguay Round of Multilateral Trade Negotiations', GATT Doc. No. MTN.TNC/W/FA, 20 December.

GATT, Uruguay Round Trade Negotiations Committee (1992a) 'Trade Negotiations Committee, Twentieth Meeting: 20 December 1991', GATT Doc. No. MTN.TNC/24, 9 January.

GATT, Uruguay Round Trade Negotiations Committee (1992b) 'Trade Negotiations Committee, Meeting on 13 January 1992, Opening Statement and concluding remarks by the Chairman', GATT Doc. No. MTN.TNC/W/99, 15 January.

GATT, Uruguay Round Trade Negotiations Committee (1992c) 'Trade Negotiations Committee, Twenty-first Meeting: 13 January 1992', GATT Doc. No. MTN.TNC/25, 5 February.

GATT, Uruguay Round Trade Negotiations Committee (1992d) 'Trade Negotiations Committee, Twenty-Second Meeting: 10 November 1992', GATT Doc. No. MTN.TNC/26, 20 November.

GATT, Uruguay Round Trade Negotiations Committee (1993a) 'Trade Negotiations Committee, Twenty-Fourth Meeting: 18 December 1992', GATT Doc. No. MTN.TNC/28, 13 January.

GATT, Uruguay Round Trade Negotiations Committee (1993b) 'Trade Negotiations Committee, Twenty-Fifth Meeting: 19 January 1993', GATT Doc. No. MTN.TNC/29, 3 February.

GATT, Uruguay Round Trade Negotiations Committee (1993c) 'Report on the Uruguay Round', GATT Doc. No. MTN.TNC/W/113, 13 July.

GATT, Uruguay Round Trade Negotiations Committee (1993d) 'Trade Negotiations Committee, Twenty-Eighth Meeting: 28 July 1993', GATT Doc. No. MTN.TNC/32, 18 August.

GATT, Uruguay Round Trade Negotiations Committee (1993e) 'Trade Negotiations Committee: Thirty-Third Meeting: 19 November 1993', GATT Doc. No. MTN.TNC/37, 29 November.

GATT, Uruguay Round Trade Negotiations Committee (1993f) 'Trade Negotiations Committee, Thirty-Fourth Meeting: 26 November 1993', GATT Doc. No. MTN.TNC/38, 2 December.

GATT, Uruguay Round Trade Negotiations Committee (1993g) 'Trade Negotiations Committee, Thirty-Sixth Meeting: 15 December 1993', GATT Doc. No. MTN.TNC/40, 21 December.

GATT, Uruguay Round Trade Negotiations Committee (1993h) 'Derestriction of the Final Act', GATT Doc. No. MTN.TNC/W/129, 21 December.

GATT, Uruguay Round Trade Negotiations Committee (1994) 'Derestriction of Certain Uruguay Round Documents: Decision of 30 March 1994', GATT Doc. No. MTN.TNC/42, 30 March.

GATT, Uruguay Round Trade Negotiations Committee (1996) 'Derestriction of Uruguay Round Documents: Documents proposed for derestriction on 3 January 1997', GATT Doc. No. MTN.TNC/46, 31 October.

GATT, Uruguay Round Trade Negotiations Committee (1997a) 'Derestriction of Uruguay Round Documents: Documents derestricted on 3 January 1997', GATT Doc. No. MTN.TNC/47, 6 January.

GATT, Uruguay Round Trade Negotiations Committee (1997b) 'Derestriction of Uruguay Round Documents: Documents derestricted on 3 January 1997: Addendum', GATT Doc. No. MTN.TNC/47/Add.1, 16 September.

Gawande, Kishore and Wendy L. Hansen (1999) 'Retaliation, Bargaining, and the Pursuit of "Free and Fair" Trade', *International Organization*, vol. 53 (Winter), pp. 117–59.

George, Alexander L. (1979) 'Case Studies and Theory Development: The Method of Structured, Focused Comparison', in P. Lauren (ed.), *Diplomacy: New Approaches in History, Theory and Policy* (New York: Free Press), pp. 43–68.

George, Aurelia (1980) 'The Strategies of Influence: Japan's Agricultural Co-operatives (*Nōkyō*) as a Pressure Group', Ph.D. dissertation, Australian National University.

George, Aurelia (1981) 'The Japanese Farm Lobby and Agricultural Policy-Making', *Pacific Affairs*, vol. 54 (Fall), pp. 409–30.

George, Aurelia (1988) *Rice Politics in Japan* (Canberra, Australia: Australia–Japan Research Centre Pacific Economic Paper No. 159, May).

George, Aurelia (1990) 'The Last Bastion: Prospects for Liberalizing Japan's Rice Market', (Tokyo: Food and Agriculture Policy Research Center (FAPRC) Japan Study Group on International Issues (SGII), Report, no. 7, pp. 113–44).

George, Aurelia and David Rapkin (1993) *GATT Negotiations and the Opening of Japan's Rice Market: A Two-Level Game Approach* (Canberra, Australia: Australia-Japan Research Centre Pacific Economic Paper No. 215, January).

Gerschenkron, Alexander (1962) 'Economic Backwardness in Historical Perspective', in Gerschenkron, *Economic Backwardness in Historical Perspective, A Book of Essays* (Cambridge, Mass.: Harvard University Press), pp. 5–30.

Gilligan, Michael J. (1997) *Empowering Exporters: Reciprocity, Delegation, and Collective Action in American Trade Policy* (Ann Arbor, Mich.: University of Michigan Press).

Gilpin, Robert (1988) 'The Implications of the Changing Trade Regime for U.S.–Japanese Relations', in Inoguchi and Okimoto (eds), *The Political Economy of Japan, Vol. 2*, pp. 138–70.

Gilpin, Robert (1989) 'The Global Context', in Iriye and Cohen (eds), *The United States and Japan in the Postwar World*, pp. 3–20.

Gilpin, Robert (1990) 'Where Does Japan Fit In?' in Newland (ed.), *The International Relations of Japan*, pp. 5–22.

Gilpin, Robert, with the assistance of Jean M. Gilpin (1987) *The Political Economy of International Relations* (Princeton, NJ: Princeton University Press).

Glick, Leslie Alan (1984) *Multilateral Trade Negotiations: World Trade after the Tokyo Round* (Totowa, NJ: Rowman & Allanheld).

Goldstein, Judith (1993) *Ideas, Interests, and American Trade Policy* (Ithaca, NY: Cornell University Press).

Goldstein, Judith (1996) 'International Law and Domestic Institutions: Reconciling North American "Unfair" Trade Laws', *International Organization*, vol. 50 (Autumn), pp. 541–64.

Goldstein, Judith and Robert O. Keohane (eds) (1993) *Ideas and Foreign Policy: Beliefs, Institutions, and Political Change* (Ithaca, NY: Cornell University Press).

Golt, Sidney (1988) *The GATT Negotiations 1986–1990: Origins, Issues & Prospects* (Washington, DC: British-North American Committee).

Goodman, David and Michael Redclift (1989) *The International Farm Crisis* (New York: St. Martin's Press).

Gordon, Peter Jegi (1990) 'Rice Policy of Japan's LDP: Domestic Trends Toward Agreement', *Asian Survey*, vol. 30 (October), pp. 943–58.

Gourevitch, Peter (1978a) 'The International System and Regime Formation: A Critical Review of Anderson and Wallerstein', *Comparative Politics* (April), pp. 419–38.

Gourevitch, Peter (1978b) 'The Second Image Reversed: The International Sources of Domestic Politics', *International Organization*, vol. 32 (Autumn), pp. 881–912.

Gourevitch, Peter (1986) *Politics in Hard Times: Comparative Responses to International Economic Crises* (Ithaca, NY: Cornell University Press).

Gourevitch, Peter (1996) 'Squaring the Circle: The Domestic Sources of International Cooperation', *International Organization*, vol. 50 (Spring) pp. 349–73.

Gourevitch, Peter, Takashi Inoguchi and Courtney Purrington (eds) (1995) *United States–Japan Relations and International Institutions After the Cold War* (La Jolla, Calif.: Graduate School of International Relations and Pacific Studies, University of California, San Diego).

Gowa, Joanne (1988) 'Public Goods and Political Institutions: Trade and Monetary Policy Processes in the United States', *International Organization*, vol. 42 (Winter) pp. 15–32.

Gowa, Joanne (1989) 'Bipolarity, Multipolarity, and Free Trade', *American Political Science Review*, vol. 83 (December) pp. 135-46.

Gowa, Joanne (1994) *Allies, Adversaries, and International Trade* (Princeton, NJ: Princeton University Press).

Graham, John L. (1993) 'The Japanese Negotiation Style: Characteristics of a Distinct Approach', *Negotiation Journal* (April), pp. 123–40.

Graham, John L. and Yoshihiro Sano (1984) *Smart Bargaining: Doing Business with the Japanese* (Cambridge, Mass.: Ballinger Publishing).

Green, Randy (2000) 'The Uruguay Round Agreement on Agriculture', *Law and Policy in International Business*, vol. 31 (Spring), pp. 819–35.

Grieco, Joseph (1988) 'Anarchy and the Limits of Cooperation: A Realist Critique of the Newest Liberal Institutionalism', *International Organization*, vol. 42 (Summer), pp. 485–507.

Grieco, Joseph (1990) *Cooperation among Nations: Europe, America, and Non-Tariff Barriers to Trade* (Ithaca, NY: Cornell University Press).

Grossman, Gene M. and Elhanan Helpman (2002) *Interest Groups and Trade Policy* (Princeton, NJ: Princeton University Press).

Guither, Harold D. (1980) *The Food Lobbyists: Behind the Scenes of Food and Agri-Politics* (Lexington, Mass.: Lexington Books).

Haas, Peter M. (1992) 'Introduction: Epistemic Communities and International Policy Coordination', *International Organization*, vol. 46 (Winter) pp. 1–35.

Haen, Hartwig de, Glenn L. Johnson and Stefan Tangermann (1985) *Agriculture and International Relations: Analysis and Policy* (London: Macmillan).

Haggard, Stephan and Beth Simmons (1987) 'Theories of International Regimes', *International Organization*, vol. 41 (Summer), pp. 491–517.

Hajnal, Peter I. (1999) *The G7/G8 System: Evolution, Role, and Documentation* (Brookfield, Vt.: Ashgate).

Haley, John (1988) 'Governance by Negotiation', in Pyle (ed.), *The Trade Crisis*, pp. 177–91; also in *Journal of Japanese Studies*, vol. 13 (Summer 1987), pp. 343–57.

Haley, John (1991) *Authority without Power* (New York: Oxford University Press).

Hall, Peter (1986) *Governing the Economy: The Politics of State Intervention in Britain and France* (New York: Oxford University Press).

Hampson, Fen, with Michael Hart (1995) *Multilateral Negotiations: Lessons from Arms Control, Trade, and the Environment* (Baltimore, Md.: Johns Hopkins University Press).

Hathaway, Dale E. (1983) 'The Internationalization of U.S. Agriculture', in Emery N. Castle and Kent A. Price (eds), *U.S. Interests and Global Natural Resources: Energy, Minerals, Food,* (Washington, DC: Resources for the Future).

Hathaway, Dale E. (1987) *Agriculture and the GATT: Rewriting the Rules* (Washington, DC: Institute for International Economics).

Hathaway, Dale E. and William M. Miner (eds) (1988) *World Agricultural Trade: Building a Consensus* (Halifax, Nova Scotia, Canada: The Institute for Research on Public Policy).

Hattori, Shinji (1990) *Gatto Nōgyō Kōshō (GATT Agriculture Negotiations)* (Tokyo: Fumin Kyōkai).

Hayami, Yujiro (1988) *Japanese Agriculture under Siege: The Political Economy of Agricultural Policies* (New York: St. Martin's Press).

Hayami, Yujiro and Saburo Yamada (eds) (1991) *The Agricultural Development of Japan: A Century's Perspective* (Tokyo: University of Tokyo Press).

Hayao, Kenji (1993) *The Japanese Prime Minister and Public Policy* (Pittsburgh, Pa.: University of Pittsburgh Press).

Hayashi, Kichiro (ed.) (1989) *The U.S.-Japanese Economic Relationship: Can It Be Improved?* (New York: New York University Press).

Held, David (1996) *Models of Democracy*, 2nd edn (Stanford, Calif.: Stanford University Press).

Hellmann, Donald C. (1988) 'Japanese Politics and Foreign Policy: Elitist Democracy Within an American Greenhouse', in Inoguchi and Okimoto (eds), *The Political Economy of Japan, Vol. 2*, pp. 345–78.

Hellmann, Donald C. (1989) 'The Imperatives for Reciprocity and Symmetry in U.S.–Japanese Economic and Defense Relations', in John H. Makin and Donald C. Hellmann (eds), *Sharing World Leadership? A New Era for America & Japan* (Washington, DC: American Enterprise Institute).

Higashi, Chikara (1983) *Japanese Trade Policy Formulation* (New York: Praeger).

Higashi, Chikara and G. Peter Lauter (1990) *The Internationalization of the Japanese Economy*, 2nd edn (Boston, Mass.: Kluwer).

Hilf, Meinhard and Ernst-Ulrich Petersmann (eds) (1993) *National Constitutions and International Economic Law, Studies in Transnational Economic Law, Vol. 8* (Boston, Mass.: Kluwer).

Hillman, Jimmye S. (1991) *Technical Barriers to Agricultural Trade* (Boulder, Col.: Westview Press).

Hillman, Jimmye and Robert A. Rothenberg (1988) *Agricultural Trade and Protection in Japan* (Brookfield, Vt.: Gower Publishing Company).

Hillman, Jimmye S., Alex F. McCalla, Timothy E. Josling and Robert A. Rothenberg, in collaboration with Yujiro Hayami, Kenzo Hemmi and Masayoshi Honma (1984) *Conflict and Comity: The State of Agricultural Trade Relations between the United States and Japan*, prepared for the US–Japan Advisory Commission (Tucson, AZ: V. of Arizona, Dept. of Agr. Econ., September).

Hirose, Michisada (1993) *Hojokin to Seikentō (Subsidies and the Ruling Party)* (Tokyo: Asahi Shimbunsha).

Hiscox, Michael J. (1999) 'The Magic Bullet? The RTAA, Institutional Reform, and Trade Liberalization', *International Organization*, vol. 53 (Autumn), pp. 669–98.

Hiscox, Michael J. (2002) *International Trade and Political Conflict: Commerce, Coalitions, and Mobility* (Princeton, NJ: Princeton University Press).

Hoare, J. E. (1994) *Japan's Treaty Ports and Foreign Settlements: The Uninvited Guests, 1858–1899* (Surrey: Curzon Press).

Hodges, Michael R., John J. Kirton and Joseph P. Daniels (eds) (1999) *The G8's Role in the New Millennium* (Brookfield, Vt.: Ashgate).

Hoekman, Bernard M. and Michel M. Kostecki (1995) *The Political Economy of the World Trading System: From GATT to WTO* (New York: Oxford University Press).

Holgerson, Karen M. (1998) *The Japan–U.S. Trade Friction Dilemma: The Role of Perception* (Brookfield, Vt.: Ashgate).

Hook, Glenn D. and Michael A. Weiner (eds) (1992) *The Internationalization of Japan* (New York: Routledge).

Hopkins, Raymond F. and Donald J. Puchala (1980) *Global Food Interdependence: Challenge to American Foreign Policy* (New York: Columbia University Press).

Houck, James P. (1982) 'Agreements and Policy in U.S.–Japanese Agricultural Trade', in Castle and Hemmi with Skillings (eds), *U.S.–Japanese Agricultural Trade Relations*, pp. 58–87.

'Houston Economic Summit Economic Declaration, July 11, 1990' (1991) in Bush, *1990, Book II – July 1 to December 31, 1990*, pp. 982–91.

Howe, Christopher (1996) *The Origins of Japanese Trade Supremacy: Development and Technology in Asia from 1540 to the Pacific War* (Chicago, Ill.: University of Chicago Press).

Hrebenar, Ronald J. and Clive S. Thomas (1995) 'The Japanese Lobby in Washington: How Different Is It?', in Cigler and Loomis (eds), *Interest Group Politics*, 4th edn, pp. 349–67.

Hrebenar, Ronald J., with contributions by Peter Berton, Akira Nakamura, and J. A. A. Stockwin (2000) *Japan's New Party System*, 3rd edn (Boulder, Col.: Westview Press).

Hudec, Robert E. (1986) 'The Legal Status of GATT in the Domestic Law of the United States', in Meinhard Hilf, Francis G. Jacobs and Ernst-Ulrich Petersmann (eds), *The European Community and GATT* (Boston, Mass.: Kluwer), pp. 187–249.

Hudec, Robert E. (1990) *The GATT Legal System and World Trade Diplomacy*, 2nd edn (Salem, NH: Butterworth).

Hudec, Robert E. (1993) *Enforcing International Trade Law: The Evolution of the Modern GATT Legal System* (Salem, NH: Butterworth).

Hufbauer, Gary, Diane T. Berliner and Kimberly Ann Elliott (1986) *Trade Protection in the United States: 31 Case Studies* (Washington, DC: Institute for International Economics).

Hufbauer, Gary, Jeffrey J. Schott and Kimberly Ann Elliott (1990) *Economic Sanctions Reconsidered: History and Current Policy*, 2nd edn (Washington, DC: Institute for International Economics).

Hunsberger, Warren (1964) *Japan and the United States in World Trade* (New York: Harper & Row).

Hunsberger, Warren (ed.) (1997) *Japan's Quest: The Search for International Role, Recognition, and Respect* (Armonk, NY: M. E. Sharpe).

Ibe, Hideo (1992) *Japan Thrice-Opened: An Analysis of Relations between Japan and the United States* (New York: Praeger).

Ikenberry, G. John (1988) 'Conclusion: An Institutional Approach to American Foreign Economic Policy', *International Organization*, vol. 42 (Winter) pp. 219–43.

Ikenberry, G. John, David A. Lake and Michael Mastanduno (eds) (1988a) *The State and American Foreign Economic Policy, International Organization*, vol. 42 (Winter).

Ikenberry, G. John, David A. Lake and Michael Mastanduno (1988b) 'Introduction: Approaches to Explaining American Foreign Economic Policy', *International Organization*, vol. 42 (Winter), pp. 1–14.

Iklé, Fred (1964) *How Nations Negotiate* (New York: Harper & Row).

Imamura, Naraomi, Shinji Hattori, Yoshio Yaguchi, Masaru Kagatsume and Keisuke Suganuma (1997) *WTO Taiseika no Shokuryō Nōgyō Senryaku (Amerika, Yōroppa, Ōsutoraria, Chūgoku, to Nihon) (Agricultural Strategy under the WTO System: A Comparative Study on U.S., EU, Australia, China, and Japan)* (Tokyo: Nōsangyoson Bunka Kyōkai).

Ingersent, K. A., A. J. Rayner and R. C. Hine (eds) (1994) *Agriculture in the Uruguay Round* (New York: St. Martin's Press).

Inoguchi, Takashi and Daniel I. Okimoto (eds) (1988) *The Political Economy of Japan, Vol. 2: The Changing International Context* (Stanford, Calif.: Stanford University Press).

Inoguchi, Takashi (1990) 'The Political Economy of Conservative Resurgence under Recession: Public Policies and Political Support in Japan, 1977–1983', in Pempel (ed.), *Uncommon Democracies*, pp. 189–225.

Inoue, Kyoko (1991) *MacArthur's Japanese Constitution: A Linguistic and Cultural Study of Its Making* (Chicago, Ill.: University of Chicago Press).

Institute for International Legal Information (ed.) (1992) *'The Dunkel Draft' from the GATT Secretariat* (Buffalo, NY: William S. Hein).

International Trade Reporter, various dates between 1984 and 1994 as cited in the Notes and listed in the Index.

International Trade Reporter's U.S. Import Weekly, various dates between 1979 and 1984 as cited in the Notes and listed in the Index.

Iriye, Akira and Warren I. Cohen (eds) (1989) *The United States and Japan in the Postwar World* (Lexington, Ky.: University Press of Kentucky).

Ishida, Takeshi and Ellis S. Krauss (eds) (1989) *Democracy in Japan* (Pittsburgh, Pa.: University of Pittsburgh Press).

Itoh, Mayumi (1998) *Globalization of Japan: Japanese Sakoku Mentality and U.S. Efforts to Open Japan* (New York: St. Martin's Press).

Iwasawa, Yūji (1995) *WTO no Funsō Shori* (*The Dispute Settlement of the World Trade Organization*) (Tokyo: Sanseidō).

Iwasawa, Yuji (1997) 'Constitutional Problems Involved in Implementing the Uruguay Round in Japan', in Jackson and Sykes (eds), *Implementing the Uruguay Round*, pp. 137–74.

Iwata, Kazumasa (1994) 'Rule Maker of World Trade: Japan's Trade Strategy and the World Trading System', in Yoichi Funabashi (ed.), *Japan's International Agenda* (New York: New York University Press), pp. 111–42.

Jackson, John H. (1969) *World Trade and the Law of GATT* (Indianapolis, Ind.: Bobbs-Merrill).

Jackson, John H. (1984) 'United States Law and Implementation of the Tokyo Round Negotiation', in Jackson, Louis and Matsushita (eds), *Implementing the Tokyo Round*, pp. 139–97.

Jackson, John H. (1988) 'Strengthening the International Legal Framework of the GATT–MTN System: Reform Proposals for the New GATT Round', in Petersmann and Hilf (eds), *The New GATT Round of Multilateral Trade Negotiations*, pp. 3–23.

Jackson, John H. (1989) *The World Trading System: Law and Policy of International Economic Relations* (Cambridge, Mass.: MIT Press).

Jackson, John H. (1990) *Restructuring the GATT System* (London: Pinter).

Jackson, John H. (1992) 'Status of Treaties in Domestic Legal Systems: A Policy Analysis', *American Journal of International Law*, vol. 86 (April), pp. 310–40.

Jackson, John H. (2000) *The Jurisprudence of GATT and the WTO: Insights on Treaty Law and Economic Relations* (New York: Cambridge University Press).

Jackson, John H. and Alan O. Sykes (eds) (1997) *Implementing the Uruguay Round* (New York: Oxford University Press).

Jackson, John H., Jean-Victor Louis and Mitsuo Matsushita (1984) *Implementing the Tokyo Round: National Constitutions and International Economic Rules* (Ann Arbor, Mich.: University of Michigan Press).

Jackson, John H., with comments by Robert O. Keohane and Gardner Patterson (1987) 'Multilateral and Bilateral Negotiating Approaches for the Conduct of U.S. Trade Policies', in Stern (ed.), *U.S. Trade Policies in a Changing World Economy*, pp. 377–412.

Jackson, John H., William J. Davey and Alan O. Sykes, Jr. (1995) *Legal Problems of International Economic Relations*, 3rd edn (St. Paul, Minn.: West Publishing).

Jain, Purnendra and Takashi Inoguchi (1997) *Japanese Politics Today: Beyond Karaoke Democracy?* (New York: St. Martin's Press).

Janow, Merit E. (1994) 'Trading with an Ally: Progress and Discontent in U.S.–Japan Trade Relations', in Curtis (ed.), *The United States, Japan, and Asia: Challenges for U.S. Policy*, pp. 53–95.

Japan, Ministry of Agriculture, Forestry and Fisheries (1987) 'Basic Direction of Agricultural Policy Toward the 21st Century', *Japan's Agricultural Review*, vol. 15 (March).

Japan, Ministry of Agriculture, Forestry and Fisheries (1989) 'Japan's Agricultural Trade', *Japan's Agricultural Review*, vol. 17 (February).

Japan, Ministry of Agriculture, Forestry and Fisheries (1992a) 'The Basic Direction of New Policies for Food, Agriculture, and Rural Areas', *Japan's Agricultural Review*, vol. 21 (November).

Japan, Ministry of Agriculture, Forestry and Fisheries (1992b) *Poketto Nōrin Suisan Tōkei (Pocket Edition Statistics for Agriculture, Forestry and Fisheries)* (Tokyo: Nōrin Tōkei Kyōkai).

Japan Agrinfo Newsletter, a newsletter published monthly by the Japan International Agricultural Council, various dates as cited in the Notes and listed in the Index.

Japan Institute of International Affairs (ed.) (1982–92) *White Papers of Japan: Annual Abstracts of Official Reports and Statistics of the Japanese Government* (Tokyo: Japan Institute of International Affairs).

Japan Times (an English-language daily newspaper in Japan), various dates as cited in the Notes and listed in the Index.

Jervis, Robert (1988) 'Realism, Game Theory, and Cooperation', *World Politics*, vol. 40 (April), pp. 317–49.

Jervis, Robert (1997) *System Effects: Complexity in Political and Social Life* (Princeton, NJ: Princeton University Press).

Johnson, Chalmers (1975) 'Who Governs? An Essay on Official Bureaucracy', *Journal of Japanese Studies*, vol. 2 (Autumn), pp. 1–28.

Johnson, Chalmers (1977) 'MITI and Japanese International Economic Policy', in Scalapino (ed.), *The Foreign Policy of Modern Japan*, pp. 227–79.

Johnson, Chalmers (1982) *MITI and the Japanese Miracle: The Growth of Industrial Policy, 1925–1975* (Stanford, Calif.: Stanford University Press).

Johnson, Chalmers (1983) 'The Internationalization of the Japanese Economy', in Harumi Befu and Hiroshi Mannari (eds), *The Challenge of Japan's Internationalization: Organization and Culture* (Nishinomiya, Hyōgo, Japan: Kwansei Gakuin University/New York: Kodansha International USA, distributed through Harper & Row), pp. 31–57.

Johnson, D. Gale (1979) 'World Agricultural and Trade Policies: Impact on U.S. Agriculture', in William Fellner (project director), *Contemporary Economic Problems 1979*, (Washington, DC: American Enterprise Institute for Public Policy Research), pp. 293–324.

Johnson, D. Gale (ed.) (1987) *Agricultural Reform Efforts in the United States and Japan* (New York: New York University Press).

Johnson, D. Gale (1991) *World Agriculture in Disarray*, 2nd edn (New York: St. Martin's Press).

Johnson, D. Gale and John A. Schnittker (eds) (1974) *U.S. Agriculture in a World Context: Policies and Approaches for the Next Decade* (New York: Praeger).

Johnson, D. Gale, Kenzo Hemmi and Pierre Lardinois (1985) *Agricultural Policy and Trade: Adjusting Domestic Programs in an International Framework, Task Force Report to the Trilateral Commission* (New York: New York University Press).

Josling, Timothy E. (1977) *Agriculture in the Tokyo Round Negotiations* (London: Trade Policy Research Centre).

Josling, Timothy E., Fred H. Sanderson and T. K. Warley (1990) 'The Future of International Agricultural Relations: Issues in the GATT Negotiations', in Sanderson (ed.), *Agricultural Protectionism in the Industrialized World*, pp. 433–64.

Josling, Timothy E., Stefan Tangermann and T. K. Warley (1996) *Agriculture in the GATT* (New York: St. Martin's Press).

Junnosuke, Masumi (1995) *Contemporary Politics in Japan* (trans. Lonny E. Carlile) (Berkeley, Calif.: University of California Press).

Karns, Margaret P. (1990) 'Multilateral Diplomacy and Trade Policy: The United States and the GATT', in Margaret P. Karns and Karen A. Mingst (eds), *The United States and Multilateral Institutions: Patterns of Changing Instrumentality and Influence* (Winchester, Mass.: Unwin Hyman), pp. 141–75.

Karube, Kensuke (1997) *Nichi-Bei Kome Kōshō, Shijō Kaihō no Shinsō to Saikōshō e no Tenbō (Japan–U.S. Rice Negotiations, The Facts Concerning Market Opening and Prospects for Renegotiation)* (Tokyo: Chūō Kōronsha).

Katzenstein, Peter J. (ed.) (1978a) *Between Power and Plenty: Foreign Economic Policies of Advanced Industrial States* (Madison, Wis.: University of Wisconsin Press).

Katzenstein, Peter J. (1978b) 'Conclusion: Domestic Structures and Strategies of Foreign Economic Policy', in Katzenstein (ed.), *Between Power and Plenty*, pp. 295–336.

Katzenstein, Peter J. (1984) *Corporatism and Change: Austria, Switzerland, and the Politics of Industry* (Ithaca, NY: Cornell University Press).

Katzenstein, Peter J. (1985) *Small States in World Markets: Industrial Policy in Europe* (Ithaca, NY: Cornell University Press).

Katzenstein, Peter J. and Yutaka Tsujinaka (1995) ' "Bullying," "Buying," and "Binding": US–Japanese Transnational Relations and Domestic Structures', in Risse-Kappen (ed.), *Bringing Transnational Relations Back In*, pp. 79–111.

Katznelson, Ira and Martin Shefter (eds) (2002) *Shaped by War and Trade: International Influences on American Political Development* (Princeton, NJ: Princeton University Press).

Kawagoe, Toshihiko (1993) 'Deregulation and Protectionism in Japanese Agriculture', in Juro Teranishi and Yutaka Kosai (eds), *The Japanese Experience of Economic Reforms* (New York: St. Martin's Press), pp. 178–204.

Keohane, Robert O. (1980) 'The Theory of Hegemonic Stability and Changes in International Economic Regimes, 1967–1977', in Ole R. Holsti, Randolph M. Siverson and Alexander L. George (eds), *Change in the International System* (Boulder, Col.: Westview Press), pp. 131–62.

Keohane, Robert O. (1984) *After Hegemony: Cooperation and Discord in the World Political Economy* (Princeton, NJ: Princeton University Press).

Keohane, Robert O. and Joseph Nye (eds) (1972) *Transnational Relations and World Politics* (Cambridge, Mass.: Harvard University Press).

Keohane, Robert O. and Nye (eds) (1989) *Power and Interdependence*, 2nd edn (Glenview, Ill.: Scott, Foresman).

Keohane, Robert O. and Helen V. Milner (eds) (1996) *Internationalization and Domestic Politics* (New York: Cambridge University Press).

Kernell, Samuel (ed.) (1991) *Parallel Politics: Economic Policymaking in Japan and the United States* (Washington, DC: Brookings Institution).

Kim, Hyung-Ki, Michio Muramatsu and T. J. Pempel (eds) (1995) *The Japanese Civil Service and Economic Development: Catalysts of Change* (New York: Oxford University Press).

Kindleberger, Charles P. (1977) 'U.S. Foreign Economic Policy, 1776–1976', *Foreign Affairs*, vol. 55 (January), pp. 395–417.

Kindleberger, Charles P. (2000) *Comparative Political Economy: A Retrospective* (Cambridge, Mass.: MIT Press).

Knoke, David, Franz Urban Pappi, Jeffrey Broadbent and Yutaka Tsujinaka (1996) *Comparing Policy Networks: Labor Politics in the U.S., Germany, and Japan* (New York: Cambridge University Press).

Knopf, Jeffrey W. (1998) *Domestic Society and International Cooperation: The Impact of Protest on US Arms Control Policy* (New York: Cambridge University Press).

Komiya, Ryutaro (1990) *The Japanese Economy: Trade, Industry, and Government* (Tokyo: University of Tokyo Press).

Komiya, Ryutaro and Motoshige Itoh (1988) 'Japan's International Trade and Trade Policy, 1955–1984', in Inoguchi and Okimoto (eds), *The Political Economy of Japan, Vol. 2*, pp. 173–224.

Komiya, Ryūtarō, Keiichi Yokobori and Tetsuo Nakata (1990) *Sekai Bōeki Taisei – Uruguai Raundo to Tsūshō Seisaku* (*The World Trading System – The Uruguay Round and Trade Policy*) (Tokyo: Tōyō Keizai Shinpōsha).

Kosaka, Masataka (1977) 'The International Economic Policy of Japan', in Scalapino (ed.), *The Foreign Policy of Modern Japan*, pp. 207–26.

Krasner, Stephen (1976) 'State Power and the Structure of International Trade', *World Politics*, vol. 28 (April), pp. 317–47.

Krasner, Stephen (1978a) 'United States Commercial and Monetary Policy: Unravelling the Paradox of External Strength and Internal Weakness', in Katzenstein (ed.), *Between Power and Plenty*, pp. 51–87.

Krasner, Stephen (1978b) *Defending the National Interest: Raw Materials Investments and U.S. Foreign Policy* (Princeton, NJ: Princeton University Press).

Krasner, Stephen (ed.) (1982a) *International Regimes* (Ithaca, NY: Cornell University Press).

Krasner, Stephen (1982b) 'Structural Causes and Regime Consequences: Regimes as Intervening Variables', in Krasner (ed.), *International Regimes*, pp. 1–21.

Krasner, Stephen (1987) *Asymmetries in Japanese–American Trade: The Case for Specific Reciprocity* (Berkeley, Calif.: University of California Institute of International Studies).

Kratochwil, Friedrich V. (1989) *Rules, Norms, and Decisions: On the Conditions of Practical and Legal Reasoning in International Relations and Domestic Affairs* (New York: Cambridge University Press).

Krauss, Ellis S. (1993) 'U.S.–Japan Negotiations on Construction and Semiconductors, 1985–1988: Building Friction and Relation-Chips', in Evans, Jacobson and Putnam (eds), *Double-Edged Diplomacy*, pp. 265–99.

Kremenyuk, Victor (ed.) (1991) *International Negotiation: Analysis, Approaches, Issues* (San Francisco, Calif.: Jossey-Bass).

Krueger, Anne O. (ed.) (1998) *The WTO as an International Organization* (Chicago, Ill.: University of Chicago Press).

Krugman, Paul (ed.) (1995) *Trade with Japan: Has the Door Opened Wider?* (Chicago, Ill.: University of Chicago Press).

Kyogoku, Jun-ichi (1987) *The Political Dynamics of Japan* (trans. Nobutaka Ike) (Tokyo: University of Tokyo Press).

LaFeber, Walter (1997) *The Clash: A History of U.S.–Japan Relations* (New York: W. W. Norton).

Lake, David A. (1988) *Power, Protection, and Free Trade: International Sources of U.S. Commercial Strategy, 1887–1939* (Ithaca, NY: Cornell University Press).

Lande, Stephen L. and Craig VanGrasstek (1986) *The Trade and Tariff Act of 1984: Trade Policy in the Reagan Administration* (Lexington, Mass.: DC Heath).

Langdon, Frank C. (1973) *Japan's Foreign Policy* (Vancouver: University of British Columbia Press).

Lavergne, Réal P. (1983) *The Political Economy of U.S. Tariffs: An Empirical Analysis* (New York: Academic Press).

Leebron, David W. (1997) 'Implementation of the Uruguay Round Results in the United States', in Jackson and Sykes (eds), *Implementing the Uruguay Round*, pp. 175–242.

Leitch, Richard D. Jr., Akira Kato and Martin Weinstein (1995) *Japan's Role in the Post-Cold War World* (Westport, Conn.: Greenwood Press).

Lenway, Stefanie Ann (1985) *The Politics of U.S. International Trade: Protection, Expansion and Escape* (Marshfield, Mass.: Pitman).

Levy, Peter (1996) *Encyclopedia of the Reagan–Bush Years* (Westport, Conn.: Greenwood Press).

Lincoln, Edward J. (1990) *Japan's Unequal Trade* (Washington, DC: Brookings Institution).

Lincoln, Edward J. (1993) *Japan's New Global Role* (Washington, DC: Brookings Institution).

Lincoln, Edward J. (1999) *Troubled Times: U.S.–Japan Trade Relations in the 1990s* (Washington, DC: Brookings Institution).

Lindblom, Charles E. (1977) *Politics and Markets: The World's Political Economic Systems* (New York: Basic Books).

Lipson, Charles (1982) 'The Transformation of Trade: The Sources and Effects of Regime Change', in Krasner (ed.), *International Regimes*, pp. 233–71.

Lipson, Charles (1984) 'International Cooperation in Economic and Security Affairs', *World Politics*, vol. 37 (October), pp. 1–23.

Lockwood, William W. (1966) 'Political Economy', in Herbert Passin (ed.), *The United States and Japan* (Englewood Cliffs, NJ: Prentice-Hall), pp. 93–127.

Lockwood, William W. (1968) *The Economic Development of Japan: Growth and Structural Change*, expanded edn (Princeton, NJ: Princeton University Press).

Lockwood, William W. (1975) 'Economic Developments and Issues', in Herbert Passin (ed.), *The United States and Japan*, 2nd edn (Washington, DC: Columbia Books), pp. 79–133.

Lohmann, Susanne and Sharyn O'Halloran (1994) 'Divided Government and U.S. Trade Policy: Theory and Evidence', *International Organization*, vol. 48 (Autumn), pp. 595–632.

'London Economic Summit Conference Declaration, June 9, 1984' (1986) in Reagan, *1984, Book* I – January 1 to June 29, 1984, pp. 830–33.

Long, Olivier (1985) *Law and its Limitations in the GATT Multilateral Trade System* (Dordrecht, Netherlands: Martinus Nijhoff).

Long, William J. (1996) 'Trade and Technology Incentives and Bilateral Cooperation', *International Studies Quarterly*, vol. 40, pp. 77–106.

Lovett, William A., Alfred E. Eckes Jr. and Richard L. Brinkman (1999) *U.S. Trade Policy: History, Theory, and the WTO* (Armonk, NY: M. E. Sharpe).

Lowi, Theodore J. (1972) 'Four Systems of Policy, Politics, and Choice', *Public Administration Review*, vol. 32 (July/August), pp. 298–310.

Lowi, Theodore J. (1979) *The End of Liberalism*, 2nd edn (New York: Norton).

Lynn, Leonard H. and Timothy J. McKeown (1988) *Organizing Business: Trade Associations in America and Japan* (Washington, DC: American Enterprise Institute).

Lynn-Jones, Sean M. and Steven E. Miller (eds) (1992) *America's Strategy in a Changing World* (Cambridge, Mass.: MIT Press).

Mainichi Daily News (an English-language daily newspaper in Japan), various dates as cited in the Notes and listed in the Index.

Mainichi Shimbun (a Japanese-language daily newspaper in Japan), various dates as cited in the Notes and listed in the Index.

Mansfield, Edward D. (1992) 'The Concentration of Capabilities and International Trade', *International Organization*, vol. 46 (Summer), pp. 731–64.

Mansfield, Edward D., Helen V. Milner and B. Peter Rosendorff (2000) 'Free to Trade: Democracies, Autocracies, and International Trade', *American Political Science Review*, vol. 94 (June), pp. 305–21.

Mansfield, Edward D., Helen V. Milner and B. Peter Rosendorff (2002) 'Why Democracies Cooperate More: Electoral Control and International Trade Agreements', *International Organization*, vol. 56 (Summer), pp. 477–513.

Manyin, Mark (1999) *Breaking the Silence: Japan's Behavior in the Tokyo and Uruguay Rounds of the GATT*, Ph.D. dissertation, Fletcher School of Law and Diplomacy.

March, James G. (1994) *A Primer on Decision Making: How Decisions Happen* (New York: Free Press).

Marlin-Bennett, Renée (1993) *Food Fights: International Regimes and the Politics of Agricultural Trade Disputes* (Langhorne, Pa.: Gordon and Breach).

Martin, Lisa L. (1992) *Coercive Cooperation: Explaining Multilateral Economic Sanctions* (Princeton, NJ: Princeton University Press).

Martin, Lisa L. (2000) *Democratic Commitments: Legislatures and International Co-operation* (Princeton, NJ: Princeton University Press).

Martin, Lisa L. and Beth Simmons (1998) 'Theories and Empirical Studies of International Institutions', *International Organization*, vol. 52 (Autumn), pp. 729–57.

Mastanduno, Michael (1992) 'Framing the Japan Problem: The Bush Administration and the Structural Impediments Initiative', *International Journal*, vol. 47 (Spring), pp. 235–64.

Matsushita, Mitsuo (1984) 'Japan and the Implementation of the Tokyo Round Results', in Jackson, Louis and Matsushita (eds), *Implementing the Tokyo Round*, pp. 77–138.

Matsushita, Mitsuo (1993a) *International Trade and Competition Law in Japan* (New York: Oxford University Press).

Matsushita, Mitsuo (1993b) 'Constitutional Framework of the Major Trade Laws in Japan: In the Context of the Uruguay Round', in Hilf and Petersmann (eds), *National Constitutions and International Economic Law*, pp. 275–97.

Matsushita, Mitsuo and T. J. Schoenbaum (1989) *Japanese International Trade and Investment Law* (Tokyo: University of Tokyo Press).

McCalla, Alex F. and Timothy E. Josling (1985) *Agricultural Policies and World Markets* (New York: Macmillan).

McCraw, Thomas, K. (ed.) (1986a) *America versus Japan: A Comparative Study of Business–Government Relations Conducted at the Harvard Business School* (Boston, Mass.: Harvard Business School Press).

McCraw, Thomas. K. (1986b) 'From Partners to Competitors: An Overview of the Period Since World War II', in McCraw (ed.), *America versus Japan*, pp. 1–33.

McGovern, Edmond (1986) *International Trade Regulation: GATT, the United States and the European Community* (Exeter, England: Globefield Press).

McKean, Margaret A. (1993) 'State Strength and the Public Interest', in Allinson and Sone (eds), pp. 72–104.

McKeown, Timothy (1984) 'Firms and Tariff Regime Change: Explaining the Demand for Protection', *World Politics*, vol. 36 (January), pp. 215–33.

McKinney, Jerome B. (1995) *Risking a Nation: U.S. Japanese Trade Failure and the Need for Political, Social, and Economic Reformation* (Lanham, Md.: University Press of America).

McMichael, Philip and Chul-Kyoo Kim (1994) 'Japanese and South Korean Agricultural Restructuring in Comparative and Global Perspective', in Philip McMichael (ed.), *The Global Restructuring of Agro-Food Systems* (Ithaca, NY: Cornell University Press), pp. 21–52.

Meeks, Philip J. (1993) 'Japan and Global Economic Hegemony', in Tsuneo Akaha and Frank Langdon (eds), *Japan in the Posthegemonic World* (Boulder, Col.: Lynne Rienner), pp. 41–67.

Meyerson, Christopher C. (1994) '"Patānkasareta Tagenshugi" Moderu no Kaizen – Gatto no Uruguai Raundo ni okeru Nihon ni yoru Nōgyō Seisaku no Kaikaku ni tsuite no Kentō' (Improving the 'Patterned Pluralist' Model: Japan and the Agricultural Policy Reform Debate of the Uruguay Round of the GATT), LLM thesis, Kyoto University Graduate School Faculty of Law.

Meyerson, Christopher C. (1996a) 'Designing and Testing a Model of Japanese Trade Policymaking: Japan's Role in the GATT Uruguay Round Agricultural Negotiations' (Paper presented at the annual meeting of the Association for Asian Studies, Honolulu, April 1996, short abstract appears in *Abstracts of the 1996 Annual Meeting of the Association for Asian Studies* (Honolulu, HI: Association for Asian Studies), p. 123).

Meyerson, Christopher C. (1996b) 'Comparing American and Japanese Trade Policymaking: The GATT Uruguay Round Agricultural Negotiations', in Allan Bird and Mitsuru Wakabayashi (eds), *Association of Japanese Business Studies Best Papers Proceedings 1996* (Nagoya, Japan: Association of Japanese Business Studies), pp. 315–31.

Meyerson, Christopher C. (1996c) 'Domestic Politics and International Relations in Trade Policymaking: The United States and Japan and the GATT Uruguay Round Agricultural Negotiations' (Paper presented at the annual meeting of the American Political Science Association, San Francisco, Calif., August 1996).

Meyerson, Christopher C. (1996d) 'Domestic Politics and International Relations in Trade Policymaking: The United States and Japan and the GATT Uruguay Round Agricultural Negotiations' (Paper presented at the International Studies Association – Japan Association of International Relations Joint Convention, Makuhari, Japan, September 1996).

Meyerson, Christopher C. (1997a) 'Japan's Evolving Role in Multilateral Trade Institutions: The GATT Uruguay Round Agricultural Negotiations' (Paper presented at the annual meeting of the Association for Asian Studies, Chicago, March 1997, short abstract appears in *Abstracts of the 1997 Annual Meeting of the Association for Asian Studies* (Chicago, Ill: Association for Asian Studies), p. 179).

Meyerson, Christopher C. (1997b) 'Domestic Politics and International Relations in American Trade Policymaking: U.S.–Japan Trade and the GATT

Uruguay Round Agricultural Negotiations' (Paper presented at the annual meeting of the International Studies Association, Toronto, Canada, March 1997).

Meyerson, Christopher C. (1997c) 'Designing and Testing a Model of Japanese Trade Policymaking: Japan's Role in the GATT Uruguay Round Agricultural Negotiations', in Terrence W. Moore and George Graen (eds), *Making Global Partnerships Work: Association of Japanese Business Studies (AJBS) Best Papers Proceedings 1997* (Washington, DC: AJBS), pp. 1–13.

Meyerson, Christopher C. (1997d) 'Domestic Politics and International Relations in American Trade Policymaking: The United States and the GATT Uruguay Round Agricultural Negotiations' (Paper presented at the annual meeting of the American Political Science Association, Washington, DC, August 1997).

Meyerson, Christopher C. (1999) 'Domestic Politics and International Relations in Trade Policymaking: The United States and Japan and the GATT Uruguay Round Agriculture Negotiations' (Paper presented at the annual meeting of the International Studies Association, Washington, DC (Columbia International Affairs Online, Columbia University Press, February 1999 – http://www.ciaonet.org/isa/mec01 (5 March 2003)).

Meyerson, Christopher C. (2000) 'Domestic Politics and International Relations in Trade Policymaking: The United States and Japan in the GATT Uruguay Round Agriculture Negotiations' (Paper presented at the annual meeting of the International Studies Association, Los Angeles, March 2000 (Columbia International Affairs Online, Columbia University Press, May 2000 – http://www.ciaonet.org/isa/mec02 (12 August 2002)).

Meyerson, Christopher C. (2002) 'Trade Policy Making in the Bush Administration: United States – Japan Trade and the GATT Uruguay Round Negotiations', in Meena Bose and Rosanna Perotti (eds), *From Cold War to New World Order: The Foreign Policy of George Bush* (Westport, Conn.: Greenwood Press).

Mikanagi, Yumiko (1996) *Japan's Trade Policy: Action or Reaction?* (New York: Routledge).

Milner, Helen V. (1987) 'Resisting the Protectionist Temptation: Industry and the Making of Trade Policy in France and the United States during the 1970s', *International Organization*, vol. 41 (Autumn), pp. 639–65.

Milner, Helen V. (1988) *Resisting Protectionism: Global Industries and the Politics of International Trade* (Princeton, NJ: Princeton University Press).

Milner, Helen V. (1992) 'International Theories of Cooperation among Nations: Strengths and Weaknesses', *World Politics*, vol. 44 (April), pp. 466–96.

Milner, Helen V. (1997) *Interests, Institutions, and Information: Domestic Politics and International Relations* (Princeton, NJ: Princeton University Press).

Milner, Helen V. (1998) 'Rationalizing Politics: The Emerging Synthesis of International, American, and Comparative Politics', *International Organization*, vol. 52 (Autumn), pp. 759–86.

Milner, Helen V. and David B. Yoffie (1989) 'Between Free Trade and Protectionism: Strategic Trade Policy and a Theory of Corporate Trade Demands', *International Organization*, vol. 43 (Spring), pp. 239–72.

Milner, Helen V. and B. Peter Rosendorff (1996) 'Trade Negotiations, Information and Domestic Politics: The Role of Domestic Groups', *Economics and Politics*, vol. 8 (July), pp. 145–89.

Milner, Helen V. and B. Peter Rosendorff. (1997) 'Democratic Politics and International Trade Negotiations: Elections and Divided Government as Constraints on Trade Liberalization', *Journal of Conflict Resolution*, vol. 41 (February), pp. 117–46.

'Ministerial Declaration' (1983) in GATT, *GATT Activities 1982* (Geneva: General Agreement on Tariffs and Trade), pp. 8–26.

Miyashita, Akitoshi and Yoichiro Sato (eds) (2001) *Japanese Foreign Policy in Asia and the Pacific: Domestic Interests, American Pressure, and Regional Integration* (New York: Palgrave).

Mnookin, Robert H., Scott R. Peppet and Andrew S. Tulumello (2000) *Beyond Winning: Negotiating to Create Value in Deals and Disputes* (Cambridge, Mass.: Harvard University Press).

Molyneux, Candido Tomas Garcia (2001) *Domestic Structures and International Trade: The Unfair Trade Instruments of the United States and the European Union* (Oxford, England: Hart).

Moon, Bruce E. (1996) *Dilemmas of International Trade* (Boulder, Col.: Westview Press).

Moore, Barrington, Jr. (1966) *Social Origins of Dictatorship and Democracy* (Boston, Mass.: Beacon Press).

Moravcsik, Andrew (1993) 'Introduction: Integrating International and Domestic Theories of International Bargaining', in Evans, Jacobson and Putnam (eds), *Double-Edged Diplomacy*, pp. 3–42.

Morishima, Masaru (ed.) (1991) *Kome Yunyū Jiyūka no Eikyō Yosoku* (*Estimated Effects of Rice Import Liberalization*) (Tokyo: Fumin Kyōkai).

Morse, Edward (1976) *Modernization and the Transformation of International Relations* (New York: Free Press).

Moyer, H. Wayne and Timothy E. Josling (1990) *Agricultural Policy Reform: Politics and Process in the EC and the USA* (Ames, Ia.: Iowa State University Press).

Mulgan, Aurelia George (1997) 'The Role of Foreign Pressure (*Gaiatsu*) in Japan's Agricultural Trade Liberalization', *The Pacific Review*, vol. 10, pp. 165–209.

Muramatsu, Michio (1993) 'Patterned Pluralism under Challenge: The Policies of the 1980s', in Allinson and Sone (eds), *Political Dynamics in Contemporary Japan*, pp. 50–71.

Muramatsu, Michio and Ellis S. Krauss (1984) 'Bureaucrats and Politicians in Policy-Making: The Case of Japan', *American Political Science Review*, vol. 78 (March), pp. 126–46.

Muramatsu, Michio and Ellis S. Krauss (1987) 'The Conservative Policy Line and the Development of Patterned Pluralism', in Yamamura and Yasuba (eds), *The Political Economy of Japan, Vol. 1: The Domestic Transformation*, pp. 516–54.

Muramatsu, Michio and Ellis S. Krauss (1990) 'The Dominant Party and Social Coalitions in Japan', in Pempel (ed.), *Uncommon Democracies*, pp. 282–305.

Muramatsu, Michio, Mitsutoshi Itō and Yutaka Tsujinaka (1986) *Sengo Nihon no Atsuryoku Dantai* (*Postwar Japanese Pressure Groups*) (Tokyo: Tōyō Keizai).

Naka, Norio (1996) *Predicting Outcomes in United States–Japan Trade Negotiations: The Political Process of the Structural Impediments Initiative* (Westport, Conn.: Quorum).

Nakakita, Toru (1993) 'Trade and Capital Liberalization Policies in Postwar Japan', in Juro Teranishi and Yutaka Kosai (eds), *The Japanese Experience of Economic Reforms* (New York: St. Martin's Press), pp. 331–65.

Nakamura, Takafusa (1997) 'Depression, Recovery, and War, 1920–1945', in Yamamura (ed.), *The Economic Emergence of Modern Japan*, pp. 116–58.

Nakano, Minoru (1997) *The Policy-Making Process in Contemporary Japan*, trans. by Jeremy Scott (New York: St. Martin's Press).

Nau, Henry R. (ed.) (1989) *Domestic Trade Politics and the Uruguay Round* (New York: Columbia University Press).

Nester, William R. (1990) *Japan's Growing Power over East Asia and the World Economy* (London: Macmillan).

Nester, William R. (1996) *Power Across the Pacific: A Diplomatic History of American Relations with Japan* (New York: New York Press).

Neu, Charles E. (1975) *The Troubled Encounter: The United States and Japan* (New York: John Wiley).

Neumann, William L. (1963) *America Encounters Japan: From Perry to MacArthur* (Baltimore, Md.: Johns Hopkins Press).

Newland, Kathleen (ed.) (1990) *The International Relations of Japan* (Basingstoke: Macmillan, in association with *Millennium*).

New York Times (newspaper), various dates as cited in the Notes and listed in the Index.

Nihon Keizai Shimbun (a Japanese-language daily newspaper in Japan), various dates as cited in the Notes and listed in the Index.

Nishio, Masaru and Michio Muramatsu (eds) (1994) *Kōza: Gyōseigaku, Seisaku to Gyōsei Daisanka* (*A Course in Public Administration: Policy and Public Administration, Vol. 3*) (Tokyo: Yūhikaku).

Nollen, Stanley D. and Dennis P. Quinn (1994) 'Free Trade, Fair Trade, Strategic Trade, and Protectionism in the U.S. Congress, 1987–88', *International Organization*, vol. 48 (Summer), pp. 491–525.

Odell, John S. (2000) *Negotiating the World Economy* (Ithaca, NY: Cornell University Press).

Odell, John S. and Thomas D. Willett (eds) (1990) *International Trade Policies: Gains from Exchange between Economics and Political Science* (Ann Arbor, Mich.: University of Michigan Press).

O'Donnell, Guillermo (1973) *Modernization and Bureaucratic-Authoritarianism* (Berkeley, Calif.: University of California Institute of International Studies).

Ogura, Kazuo (1983) *Trade Conflict: A View from Japan* (Washington, DC: Japan Economic Institute).

Ogura, Takekazu B. (1989) 'A Preliminary Approach to Reforming Japanese Agricultural Structure', (Tokyo: Food and Agriculture Policy Research Center (FAPRC) Japan SGII Report, no. 3, November).

O'Halloran, Sharyn (1994) *Politics, Process, and American Trade Policy* (Ann Arbor, Mich.: University of Michigan Press).

Okimoto, Daniel (ed.) (1982) *Japan's Economy: Coping with Change in the International Environment* (Boulder, Col.: Westview Press).

Okimoto, Daniel (1988) 'Political Inclusivity: The Domestic Structure of Trade', in Inoguchi and Okimoto (eds), pp. 305–44.

Okimoto, Daniel (1989) *Between MITI and the Market: Japanese Industrial Policy for High Technology* (Stanford, Calif.: Stanford University Press).

Organisation for Economic Co-operation and Development (1987) *National Policies and Agricultural Trade: Country Study, Japan* (Paris: OECD).

Organisation for Economic Co-operation and Development (1993) *Assessing the Effects of the Uruguay Round* (Paris: OECD).

Organisation for Economic Co-operation and Development (1995) *The Uruguay Round: A Preliminary Evaluation of the Impacts of the Agreement on Agriculture in the OECD Countries* (Paris: OECD).

Organisation for Economic Co-operation and Development (1997) *The Uruguay Round Agreement on Agriculture and Processed Agricultural Products* (Paris: OECD).

Otsuka, Keijiro and Yujiro Hayami (1984) 'Rice Policy in Japan: Its Costs and Distributional Consequences' (Canberra, Australia: Australia–Japan Research Centre Research Paper No. 114, August).

Ouchi, Tsutomu (1989) 'Agriculture Policy Reform Why Now?' (Tokyo: Food and Agriculture Policy Research Center (FAPRC) Japan Study Group on International Issues (SGII) Report, no. 2, October).

Ōuchi, Tsutomu and Naomi Saeki (eds) (1991) *Gatto Nōgyō Kōshō to Nihon Nōgyō (GATT Agriculture Negotiations and Japanese Agriculture)* (Tokyo: Nōrin Tōkei Kyōkai).

Oye, Kenneth A. (ed.) (1986a) *Cooperation under Anarchy* (Princeton, NJ: Princeton University Press).

Oye, Kenneth A. (1986b) 'Explaining Cooperation Under Anarchy: Hypotheses and Strategies', in Oye (ed.), *Cooperation Under Anarchy*, pp. 1–24.

Paarlberg, Don (1980) *Farm and Food Policy: Issues of the 1980s* (Lincoln, Nebr.: University of Nebraska Press).

Paarlberg, Philip L. and J. A. Sharples (1984) *Japanese and European Community Agricultural Trade Policies: Some US Strategies* (Washington, DC: USDA Foreign Agricultural Economic Report No. 204).

Paarlberg, Robert L. (1988) *Fixing Farm Trade: Policy Options for the United States* (Cambridge, Mass.: Ballinger Publishing).

Paarlberg, Robert L. (1993) 'Why Agriculture Blocked the Uruguay Round: Evolving Strategies in a Two-Level Game', in Avery (ed.), *World Agriculture and the GATT*, pp. 39–54.

Paarlberg, Robert L. (1997) 'Agricultural Policy Reform and the Uruguay Round: Synergistic Linkage in a Two-Level Game?', *International Organization*, vol. 51 (Summer), pp. 413–44.

Pasour, E. C., Jr. (1990) *Agriculture and the State: Market Processes and Bureaucracy* (New York: Holmes and Meier).

Pastor, Robert A. (1980) *Congress and the Politics of U.S. Foreign Economic Policy, 1929–1976* (Berkeley, Calif.: University of California Press).

Pastor, Robert A. (1983) 'The Cry-and-Sigh Syndrome: Congress and Trade Policy', in Allen Schick (ed.), *Making Economic Policy in Congress* (Washington, DC: American Enterprise Institute), pp. 158–95.

Patrick, Hugh (ed.) (1976) *Japanese Industrialization and its Social Consequences* (Berkeley, Calif.: University of California Press).

Patrick, Hugh and Henry Rosovsky (eds) (1976) *Asia's New Giant: How the Japanese Economy Works* (Washington, DC: Brookings Institution).

Patrick, Hugh and Hideo Sato (1982) 'The Political Economy of United States–Japan Trade in Steel', in Yamamura (ed.), *Policy and Trade Issues of the Japanese Economy*, pp. 197–238.

Pempel, T. J. (ed.) (1977) *Policymaking in Contemporary Japan* (Ithaca, NY: Cornell University Press).

Pempel, T. J. (1978) 'Japanese Foreign Economic Policy: The Domestic Bases for International Behavior', in Katzenstein (ed.), *Between Power and Plenty*, pp. 139–90.

Pempel, T. J. (1987) 'The Unbundling of "Japan, Inc.": The Changing Dynamics of Japanese Policy Formation', in Pyle (ed.), *The Trade Crisis*, pp. 117–52.

Pempel, T. J. (ed.) (1990) *Uncommon Democracies: The One-Party Dominant Regimes* (Ithaca, NY: Cornell University Press).

Pempel, T. J. (1993) 'From Exporter to Investor: Japanese Foreign Economic Policy', in Curtis (ed.), *Japan's Foreign Policy After the Cold War*, pp. 105–36.

Pempel, T. J. (1998) *Regime Shift: Comparative Dynamics of the Japanese Political Economy* (Ithaca, NY: Cornell University Press).

Pepper, Thomas, Merit E. Janow and Jimmy W. Wheeler (1985) *The Competition: Dealing with Japan* (New York: Praeger).

Petersmann, Ernst-Ulrich and Meinhard Hilf (1988) *The New GATT Round of Multilateral Trade Negotiations: Legal and Economic Problems* (Boston, Mass.: Kluwer).

Porges, Amelia (1994) 'Japan: Beef and Citrus', in Bayard and Elliott (eds), *Reciprocity and Retaliation in U.S. Trade Policy*, pp. 233–66.

Port, Kenneth L. (1996) *Comparative Law: Law and the Legal Process in Japan* (Durham, NC: Carolina Academic Press).

Porter, Jane M. and Douglas E. Bowers (1989) *A Short History of U.S. Agricultural Trade Negotiations* (Washington, DC: USDA Economic Research Service Staff Report No. AGES 89–23).

Preeg, Ernest H. (1970) *Traders and Diplomats: An Analysis of the Kennedy Round of Negotiations under the General Agreement on Tariffs and Trade* (Washington, DC: Brookings Institution).

Preeg, Ernest H. (1995) *Traders in a Brave New World: The Uruguay Round and the Future of the International Trading System* (Chicago, Ill.: University of Chicago Press).

President. Proclamation (1990) 'Modification of Tariffs and Quota on Certain Sugars, Syrups, and Molasses, Proclamation 6179', *Federal Register*, 55, 13 September, p. 38,293.

Prestowitz, Clyde V., Jr. (1988) *Trading Places: How We Allowed Japan to Take the Lead* (New York: Basic Books).

Pugel, Thomas A., with Robert G. Hawkins (eds) (1986) *Fragile Interdependence: Economic Issues in U.S.–Japanese Trade and Investment* (Lexington, Mass.: Lexington Books).

Pugel, Thomas A. (1987) 'Limits of Trade Policy Toward High Technology Industries: The Case of Semiconductors', in Ryuzo Sato and Paul Wachtel (eds), *Trade Friction and Economic Policy: Problems and Prospects for Japan and the United States* (New York: Cambridge University Press), pp. 184–223.

'Punta del Este Declaration, Ministerial Declaration of 20 September 1986' (1987) in *GATT Activities 1986, An Annual Review of the Work of the GATT* (Geneva: General Agreement on Tariffs and Trade), pp. 15–27.

Putnam, Robert D. (1988) 'Diplomacy and Domestic Politics: The Logic of Two-Level Games', *International Organization*, vol. 42 (Summer), pp. 427–60.

Putnam, Robert D. and Nicholas Bayne (1987) *Hanging Together: Cooperation and Conflict in the Seven-Power Summits*, revised and enlarged edn (Cambridge, Mass.: Harvard University Press).

Pyle, Kenneth B. (ed.) (1987) *The Trade Crisis: How Will Japan Respond?* (Seattle, Wash.: The Society for Japanese Studies).

Pyle, Kenneth B. (1996) *The Japanese Question: Power and Purpose in a New Era*, 2nd edn (Washington, DC: American Enterprise Institute Press).

Ragin, Charles C. (1987) *The Comparative Method: Moving Beyond Qualitative and Quantitative Strategies* (Berkeley, Calif.: University of California Press).

Raiffa, Howard (1982) *The Art and Science of Negotiation* (Cambridge, Mass.: Harvard University Press).

Ramseyer, J. Mark and Frances M. Rosenbluth (1993) *Japan's Political Marketplace* (Cambridge, Mass.: Harvard University Press).

Ramseyer, J. Mark and Frances M. Rosenbluth (1995) *The Politics of Oligarchy: Institutional Choice in Imperial Japan* (New York: Cambridge University Press).

Rapkin, David P. (ed.) (1990) *World Leadership and Hegemony: International Political Economy Yearbook, Volume 5* (Boulder, Col.: Lynne Rienner).

Rapkin, David and Aurelia George (1993) 'Rice Liberalization and Japan's Role in the Uruguay Round: A Two-Level Game Approach', in Avery (ed.), *World Agriculture and the GATT*, pp. 55–94.

Rapkin, David and William P. Avery (eds) (1995) *National Competitiveness in a Global Economy: International Political Economy Yearbook, Volume 8:* (Boulder, Col.: Lynne Rienner).

Reagan, Ronald (1983, 1984, 1985, 1986, 1988, 1989, 1990, 1991) *Public Papers of the Presidents of the United States: Ronald Reagan* (Washington, DC: Government Printing Office).

Reams, Bernard, D. Jr., and Jon S. Schultz (1995) *Uruguay Round Agreements Act: A Legislative History of Public Law No. 103–465* (Buffalo, NY: William S. Hein).

Reich, Michael R., Yasuo Endo and C. Peter Timmer (1986) 'Agriculture: The Political Economy of Structural Change', in McCraw (ed.), *America versus Japan*, pp. 151–92.

Reischauer, Edwin O. (1965) *The United States and Japan*, 3rd edn (Cambridge, Mass.: Harvard University Press).

Reischauer, Edwin O. and Marius B. Jansen (1995) *The Japanese Today: Change and Continuity*, enlarged edn (Cambridge, Mass.: Harvard University Press).

'Remarks at a White House Meeting with Business and Trade Leaders, September 23, 1985', in Reagan, *1985, Book II – June 29 to December 31, 1985*, 1988, pp. 1127–30.

'Remarks of the President and Prime Minister Yasuhiro Nakasone of Japan Following Their Meetings in Tokyo, November 10, 1983', in Reagan, *1983, Book II – July 2 to December 31, 1983*, 1985, pp. 1566–70.

'Remarks to the National Chamber Foundation, November 17, 1988', in Reagan, *1988–89, Book II – July 2, 1988 to January 19, 1989*, 1991, pp. 1521–4.

Rhodes, Carolyn (1993) *Reciprocity, U.S. Trade Policy, and the GATT Trade Regime* (Ithaca, NY: Cornell University Press).

Richardson, Bradley (1997) *Japanese Democracy: Power, Coordination, and Performance* (New Haven, Conn.: Yale University Press).

Risse-Kappen, Thomas (ed.) (1995) *Bringing Transnational Relations Back In: Non-State Actors, Domestic Structures and International Institutions* (New York: Cambridge University Press).

Rogowski, Ronald (1987) 'Trade and the Variety of Democratic Institutions', *International Organization*, vol. 41 (Spring), pp. 203–23.

Rogowski, Ronald (1989) *Commerce and Coalitions: How Trade Affects Domestic Political Alignments* (Princeton, NJ: Princeton University Press).

Rosecrance, Richard (1986) *The Rise of the Trading State: Commerce and Conquest in the Modern World* (New York: Basic Books).

Rosecrance, Richard and Arthur A. Stein (eds) (1993) *The Domestic Bases of Grand Strategy* (Ithaca, NY: Cornell University Press).

Rosenau, James N. (1997) *Along the Domestic–Foreign Frontier: Exploring Governance in a Turbulent World* (New York: Cambridge University Press).

Rosenbluth, Frances McCall (1996) 'Internationalization and Electoral Politics in Japan', in Keohane and Milner (eds), *Internationalization and Domestic Politics*, pp. 137–56.

Rothacher, Albrecht (1989) *Japan's Agro-Food Sector: The Politics and Economics of Excess Production* (New York: St. Martin's Press).

Rowley, Charles K. and Willem Thorbecke (1993) 'The Role of the Congress and the Executive in US Trade Policy Determination: A Public Choice Analysis', in Hilf and Petersmann (eds), pp. 347–69.

Ruggie, John Gerard (1982) 'International Regimes, Transactions, and Change: Embedded Liberalism in the Postwar Economic Order', *International Organization*, vol. 36 (Spring), pp. 379–415.

Ruggie, John Gerard (ed.) (1993) *Multilateralism Matters: The Theory and Praxis of an Institutional Form* (New York: Columbia University Press).

Saeki, Naomi (1990) *Gatto to Nihon Nōgyō* (*GATT and Japanese Agriculture*) (Tokyo: University of Tokyo Press).

Samuels, Richard (1987) *The Business of the Japanese State: Energy Markets in Comparative and Historical Perspective* (Ithaca, NY: Cornell University Press).

Sanderson, Fred H. (ed.) (1990) *Agricultural Protectionism in the Industrialized World* (Washington, DC: Resources for the Future).

Sartori, Giovanni (1976) *Parties and Party Systems* (Cambridge University Press).

Sato, Hideo (1996) 'The Changing International System and Trade-Conflict Management between Japan and the United States', *Journal of International Political Economy*, vol. 1 (March), pp. 7–32.

Sato, Hideo and Timothy J. Curran (1982) 'Agricultural Trade: The Case of Beef and Citrus', in Destler and Sato (eds), *Coping with U.S.–Japanese Economic Conflicts*, pp. 121–83.

Sato, Hideo and Michael W. Hodin (1982) 'The U.S.–Japanese Steel Issue of 1977', in Destler and Sato (eds), *Coping with U.S.–Japanese Economic Conflicts*, pp. 27–72.

Sato, Hideo and Gunther Schmitt (1993) 'The Political Management of Agriculture in Japan and West Germany', in Haruhiro Fukui, Peter H. Merkl, Hubertus Müller-Groeling and Akio Watanabe (eds), *The Politics of Economic Change in Postwar Japan and West Germany, Vol. 1: Macroeconomic Conditions and Policy Responses* (London: Macmillan), pp. 233–80.

Sato, Kazuo (ed.) (1999) *The Transformation of the Japanese Economy* (Armonk, NY: M. E. Sharpe).

Sato, Ryuzo (1994) *The Chrysanthemum and the Eagle: The Future of U.S.–Japan Relations* (New York: New York University Press).

Sato, Ryuzo and Paul Wachtel (eds) (1987) *Trade Friction and Economic Policy: Problems and Prospects for Japan and the United States* (New York: Cambridge University Press).

Sato, Ryuzo and Julianne Nelson (eds) (1989) *Beyond Trade Friction: Japan–U.S. Economic Relations* (New York: Cambridge University Press).

Sato, Yoichiro (1996) 'Sticky Efforts: Japan's Rice Market Opening and U.S.–Japan Transnational Lobbying', in Harumi Befu (ed.), *Japan Engaging the World: A Century of International Encounter* (Denver, Col.: Center for Japan Studies at Teikyo Loretto Heights University), pp. 73–99.

Saxonhouse, Gary and Hugh Patrick (1976) 'Japan and the United States: Bilateral Tensions and Multilateral Issues in the Economic Relationship', in Donald C. Hellmann (ed.), *China and Japan: A New Balance of Power, Critical Choices for Americans, Vol. xii* (Lexington, Mass.: Lexington Books), pp. 95–157.

Scalapino, Robert (ed.) (1977) *The Foreign Policy of Modern Japan* (Berkeley, Calif.: University of California Press).

Schattschneider, E. E. (1935) *Politics, Pressures and the Tariff: A Study of Free Private Enterprise in Pressure Politics, as Shown in the 1929–30 Revision of the Tariff* (New York: Prentice-Hall).

Schlesinger, Jacob M. (1997) *Shadow Shoguns: The Rise and Fall of Japan's Postwar Political Machine* (New York: Simon & Schuster).

Schmiegelow, Michèle (ed.) (1986) *Japan's Response to Crisis and Change in the World Economy* (Armonk, NY: M. E. Sharpe).

Schmitz, Andrew, Garth Coffin and Kenneth A. Rosaasen (eds) (1996) *Regulation and Protectionism under GATT: Case Studies in North American Agriculture* (Boulder, Col.: Westview Press).

Schodt, Frederik L. (1994) *America and the Four Japans: Friend, Foe, Model, Mirror* (Berkeley, Calif.: Stone Bridge Press).

Schoppa, Leonard J. (1993) 'Two-Level Games and Bargaining Outcomes: Why *Gaiatsu* Succeeds in Japan in Some Cases But Not Others', *International Organization*, vol. 47 (Summer), pp. 353–86.

Schoppa, Leonard J. (1997) *Bargaining with Japan: What American Pressure Can and Cannot Do* (New York: Columbia University Press).

Schoppa, Leonard J. (1999) 'The Social Context in Coercive International Bargaining', *International Organization*, vol. 53 (Spring), pp. 307–42.

Schott, Jeffrey J. (1994) *The Uruguay Round: An Assessment* (Washington, DC: Institute for International Economics).

Schwartz, Frank (1993) 'Of Fairy Cloaks and Familiar Talks: The Politics of Consultation', in Allinson and Sone (eds), pp. 217–41.

Schwartz, Frank (1998) *Advice and Consent: The Politics of Consultation in Japan* (New York: Cambridge University Press).

Searight, Amy E. (2002) 'International Organizations', in Vogel (ed.), *U.S.–Japan Relations in a Changing World*, pp. 160–97.

'Second "Maekawa" Report: Summary Report of the Economic Council's Special Committee on Economic Restructuring, April 23, 1987' (1988) trans. the Embassy of Japan, Washington, DC, in Choy, pp. 14–18.

Sheingate, Adam D. (2001) *The Rise of the Agricultural Welfare State: Institutions and Interest Group Power in the United States, France, and Japan* (Princeton, NJ: Princeton University Press).

Shoch, James (2001) *Trading Blows: Party Competition and U.S. Trade Policy in a Globalizing Era* (Chapel Hill, NC: University of North Carolina Press).

Shokuryō Seisaku Kenkyūkai (ed.) (1999) *WTO Taiseika no Kome to Shokuryō* (*Rice and Food under the WTO System*) (Tokyo: Nihon Keizai Hyōronsha).

Simmons, Beth A. (1994) *Who Adjusts? Domestic Sources of Foreign Economic Policy During the Interwar Years* (Princeton, NJ: Princeton University Press).

Smith, Patrick (1991) 'Letter from Tokyo', *The New Yorker*, 14 October, pp. 105–18.

Snidal, Duncan (1985) 'The Limits of Hegemonic Stability Theory', *International Organization*, vol. 39 (Autumn), pp. 579–614.

Snidal, Duncan (1991) 'Relative Gains and the Pattern of International Cooperation', *American Political Science Review*, vol. 85 (September), pp. 701–26.

Snyder, Jack (1991) *Myths of Empire: Domestic Politics and International Ambition* (Ithaca, NY: Cornell University Press).

Spero, Joan Edelman and Jeffrey A. Hart (1997) *The Politics of International Economic Relations*, 5th edn (New York: St. Martin's Press).

Statement of Administrative Action of the Uruguay Round Agreements Act concerning the 'Agreement on Agriculture', in US Congress, House, Message from the President of the United States Transmitting the Uruguay Round Trade Agreements, Vol. 1, 1994, pp. 709–63.

Stein, Arthur A. (1984) 'The Hegemon's Dilemma: Great Britain, the United States, and the International Economic Order', *International Organization*, vol. 38 (Spring), pp. 355–86.

Stein, Arthur A. (1990) *Why Nations Cooperate: Circumstance and Choice in International Relations* (Ithaca, NY: Cornell University Press).

Stern, Robert M. (ed.) (1987) *U.S. Trade Policies in a Changing World Economy* (Cambridge, Mass.: MIT Press).

Stewart, Terence P. (ed.) (1993) *The GATT Uruguay Round, A Negotiating History (1986–1992)*, 3 vols (Boston, Mass.: Kluwer).

Stewart, Terence P. (ed.) (1999) *The GATT Uruguay Round, A Negotiating History (1986–1994), Vol. IV: The End Game (Part I)* (Boston, Mass.: Kluwer).

Suleiman, Ezra (ed.) (1984) *Bureaucrats and Policy Making* (New York: Holmes and Meier).

Sumner, Daniel A. (1995a) *Agricultural Trade Policy: Letting Markets Work* (Washington, DC: American Enterprise Institute Press).

Sumner, Daniel A. (ed.) (1995b) *Agricultural Policy Reform in the United States* (Washington, DC: American Enterprise Institute Press).

Teranishi, Juro and Yutaka Kosai (eds) (1993) *The Japanese Experience of Economic Reforms* (New York: St. Martin's Press).

'Text of Economic Declaration by 7 Leading Industrial Nations' (1987) in *New York Times*, 11 June, sec. A, p. 16.

Thompson, Kenneth, W. (ed.) (1997) *The Reagan Presidency: Ten Intimate Perspectives of Ronald Reagan* (Lanham, Md.: University Press of America).

Thompson, Kenneth W. (ed.) (1997) *The Bush Presidency: Ten Intimate Perspectives of George Bush* (Lanham, Md.: University Press of America).

Tiedemann, Arthur E. (1974) 'Japan's Economic Foreign Policies, 1868–1893', in James William Morley (ed.), *Japan's Foreign Policy, 1868–1941: A Research Guide* (New York: Columbia University Press), pp. 118–52.

Tilton, Mark (1996) *Restrained Trade: Cartels in Japan's Basic Materials Industries* (Ithaca, NY: Cornell University Press).

'Toronto Economic Summit Conference Economic Declaration, June 21, 1988' (1990) in Reagan, *1988, Book 1 – January 1 to July 1, 1988* (Washington, DC: Government Printing Office), pp. 798–804.

Treverton, Gregory F. (1994) *Making American Foreign Policy* (Englewood Cliffs, NJ: Prentice Hall).

Tsuchiya, Keizō and Keiji Ōga (eds) (1988) *Kome no Kokusai Jukyū to Yunyū Jiyūka Mondai* (*The International Demand and Supply of Rice and the Import Liberalization Problem*) (Tokyo: Nōrin Tōkei Kyōkai).

Tweeten, Luther, Cynthia L. Dishon, Wen S. Chern, Naraomi Imamura and Masaru Morishima (eds) (1993) *Japanese and American Agriculture: Tradition and Progress in Conflict* (Boulder, Col.: Westview Press).

Unah, Isaac (1998) *The Courts of International Trade: Judicial Specialization, Expertise, and Bureaucratic Policy-Making* (Ann Arbor, Mich.: University of Michigan Press).

United States (US) Agriculture Technical Advisory Committee (1994) *The Uruguay Round of Multilateral Trade Negotiations: Report of the Agriculture Technical Advisory Committee* (Washington, DC: Government Printing Office, January).

US Code (1988), vol. 7, sec. 624, 'Limitations on imports; authority of President'.

US Congress, Congressional Budget Office (1987) *The GATT Negotiations and U.S. Trade Policy* (Washington, DC: Government Printing Office, June).

US Congress, House (1994) Message from the President of the United States Transmitting the Uruguay Round Trade Agreements, Texts of Agreements, Implementing Bill, Statement of Administrative Action and Required Supporting Statements, 103d Cong., 2d sess., H. Doc. 316, Vols I–II.

US Congress, House Committee on Agriculture (1985) *The Food Security Act of 1985*, 99th Cong., H. Rept. 271(I).

US Congress, House Committee on Agriculture (1986) *Agricultural Provision Proposals to Omnibus Trade Legislation: Hearing*, 99th Cong., 2nd sess., 15 April.

US Congress, House Committee on Agriculture, Subcommittee on Department Operations, Research, and Foreign Agriculture (1986) *Hearing to Receive GAO Reports on U.S. Department of Agriculture's Agricultural Trade Programs; and Summary of Trade Sessions Held in Uruguay*, 99th Cong., 2nd sess., 29 September.

US Congress, House Committee on Agriculture, Subcommittee on Cotton, Rice, and Sugar (1986) *Review of Japan's Policy Concerning the Importation of Rice, Including a Petition Filed by the U.S. Rice Millers' Association: Hearing*, 99th Cong., 2d sess., 1 October.

US Congress, House Committee on Agriculture, Subcommittee on Livestock, Dairy, and Poultry (1987) *Review of the Dairy Termination Program and Other Ongoing Dairy Program Initiatives Mandated in the Food Security Act of 1985; and Current and Proposed U.S. Trade Policies and their Effect on the Competitiveness of the Domestic Livestock, Dairy, and Poultry Industries: Hearing*, 100th Cong., 1st sess., 4 March.

US Congress, House Committee on Agriculture (1987) *Trade and International Economic Policy Reform Act of 1987*, 100th Cong., 1st sess., H. Rept. 40, 7 April.

US Congress, House Committee on Agriculture, Subcommittee on Department Operations, Research, and Foreign Agriculture (1991) *Review of the Uruguay Round of Multilateral Trade Negotiations under the General Agreement on Tariffs and Trade: Hearing*, 102nd Cong., 1st sess., 28 February.

US Congress, House Committee on Agriculture (1991) *Review of Fast-Track Extension Request Submitted by the Administration: Hearing*, 102nd Cong., 1st sess., 13 March.

US Congress, House Committee on Agriculture (1991, 1992) *Review of International Trade Negotiations Affecting U.S. Agricultural Policy under the General Agreement on Tariffs and Trade [GATT]: Hearings*, 102nd Cong., 1st and 2nd sess., 10 December 1991, 9 January, 25 February and 31 March 1992.

US Congress, House Committee on Agriculture (1993) *Review of the President's Supplemental Agreements to the North American Free-Trade Agreement and an Update on the Uruguay Round of the GATT Negotiations: Hearing*, 103rd Cong., 1st sess., 17 March.

US Congress, House Committee on Agriculture (1994) *Review of the Uruguay Round GATT Agreement Implications for Agricultural Trade: Hearings*, 103rd Cong., 2d sess., 16 March and 20 April.

US Congress, House Committee on Appropriations, Subcommittee on the Departments of Commerce, Justice, and State, the Judiciary, and Related Agencies (1985) *Departments of Commerce, Justice, and State, the Judiciary, and Related Agencies Appropriations for 1986*, 99th Cong., 1st sess., 27 February.

US Congress, House Committee on Foreign Affairs, Subcommittee on Europe and the Middle East (1986) *United States–European Community Trade Relations: Problems and Prospects for Resolution: Hearing*, 99th Cong,. 2nd sess., 24 July.

US Congress, House Committee on Ways and Means, Subcommittee on Foreign Trade Policy (1957) *Compendium of Papers on United States Foreign Trade Policy* (Washington, DC: Government Printing Office).

US Congress, House Committee on Ways and Means, Subcommittee on Trade (1989) *Uruguay Round of Multilateral Trade Negotiations*, 101st Cong., 1st sess., 11 April.

US Congress, House Committee on Ways and Means, Subcommittee on Trade (1993) *President's Request for Extension of 'Fast-Track' Procedures for Uruguay Round Implementation and Possible Administration Requests for Extensions of Expiring Trade Programs: Hearings*, 103rd Cong., 1st sess., 27 April.

US Congress, House Committee on Ways and Means, Subcommittee on Trade (1993) *United States–Japan Trade, Commercial, and Economic Relations: Hearing*, 103rd Cong., 1st sess., 13 July.

US Congress, House Committee on Ways and Means, Subcommittee on Trade (1993) *Uruguay Round of Multilateral Trade Negotiations*, 103rd Cong., 1st sess., 4–5 November.

US Congress, House Committee on Ways and Means, Subcommittee on Trade (1994) *Uruguay Round Agreements Act*, 103rd Cong,. 1st sess., H. Rept. 826.

US Congress, House Committee on Ways and Means (1997) *Overview and Compilation of U.S. Trade Statutes*, 105th Cong,. 1st sess., 25 June.

US Congress, House Conference Report (1988) *Omnibus Trade and Competitiveness Act of 1988*, 100th Cong, H. Rept. 576.

US Congress, Joint Economic Committee, Subcommittee on Agriculture and Transportation (1981) *The Importance of Agriculture to the U.S. Economy*, 97th Cong., 1st sess., 14-15 September.

US Congress, Joint Economic Committee (1983) *The 1983 Economic Report of the President: Hearings*, 98th Cong., 1st sess., 3 February.

US Congress, Senate Committee on Agriculture, Nutrition, and Forestry, Subcommittee on Foreign Agricultural Policy (1986) *Preparing for the GATT: A Review of Agricultural Trade Issues: Hearings*, 99th Cong., 2nd sess., 17 June.

US Congress, Senate Committee on Agriculture, Nutrition, and Forestry, Subcommittee on Domestic and Foreign Marketing and Product Promotion (1989) *General Agreement on Tariffs and Trade Proceedings in Geneva as They Relate to Agriculture: Hearing*, 101st Cong., 1st sess., 12 April.

US Congress, Senate Committee on Agriculture, Nutrition, and Forestry, Subcommittee on Agricultural Production and Stabilization of Prices (1990) *Preparation for the 1990 Farm Bill: Hearings*, 101st Cong., 2nd sess., 5 March.

US Congress, Senate Committee on Agriculture, Nutrition, and Forestry (1991) *Fast Track Procedures for Agricultural Trade Negotiations: Hearing*, 102nd Cong., 1st sess., 8 May.

US Congress, Senate Committee on Agriculture, Nutrition, and Forestry (1994) *The GATT Agreement: The Implementation of the Uruguay Round as It Affects United States Agriculture: Hearing*, 103rd Cong., 2d sess., 20 April.

US Congress, Senate Committee on Banking, Housing, and Urban Affairs, Subcommittee on International Finance and Monetary Policy (1982) *Foreign Barriers to U.S. Trade: Hearing, Part II Merchandise Exports*, 97th Cong., 2nd sess., 4 March.

US Congress, Senate Committee on Finance, Subcommittee on International Trade (1982) *U.S. Approach to 1982 Meeting of World Trade Ministers on the GATT: Hearing*, 97th Cong., 2nd sess., 1 March.

US Congress, Senate Committee on Finance (1989) *Oversight of the Trade Act of 1988: Hearing*, 101st Cong., 1st sess., 20 April.

US Congress, Senate Committee on Finance, Subcommittee on International Trade (1989) *Extending International Trading Rules to Agriculture: Hearing*, 101st Cong., 1st sess., 3 November.

US Congress, Senate Committee on Finance (1993) *Anticipated Nomination of Mickey Kantor: Hearing*, 103rd Cong., 1st sess., 19 January.

US Congress, Senate Committee on Finance (1993) *Renewal of Fast-Track Authority and the Generalized System of Preferences Program: Hearing*, 103rd Cong., 1st sess., 20 May.

US Congress, Senate Committee on Finance, Committee on Agriculture, Nutrition, and Forestry, Committee on Governmental Affairs (1994) *Uruguay Round Agreements Act: Joint Report to Accompany S. 2467, Uruguay Round Agreements Act*, 103d Cong., 22 November, 2d sess., S. Rept. 412.

US Constitution, Art. I, §8.

US Department of Agriculture (1990) *Multilateral Trade Reform: What the GATT Negotiations Mean to U.S. Agriculture* (Washington, DC: USDA Staff Briefing, August).

US Department of Agriculture, Economic Research Service (1987) *Government Intervention in Agriculture: Measurement, Evaluation, and Implications for Trade Negotiations* (Washington, DC: FAER-229, April).

US Department of Agriculture, Economic Research Service (1988) *Agricultural Policy Reform in the Uruguay Round: Proceedings of a Workshop on Economic Issues and Research Needs* (Washington, DC: Staff Report No. AGES-880729, September).

US Department of Agriculture, Economic Research Service (1989) *Agricultural-Food Policy Review: U.S. Agricultural Policies in a Changing World* (Washington, DC: USDA Agricultural Economic Report No. 620, November).

US Department of Agriculture, Economic Research Service (1989) *Bibliography of Research Supporting the Uruguay Round of the GATT* (Washington, DC: Staff Report No. AGES 89–64).

US Department of Agriculture, Economic Research Service (1990) *The Basic Mechanisms of Japanese Farm Policy* (Washington, DC: US Department of Agriculture, Economic Research Service Miscellaneous Publication No. 1478, February).

US Department of Agriculture, Economic Research Service (1994) *Effects of the Uruguay Round Agreement on U.S. Agricultural Commodities* (Washington, DC: US Department of Agriculture, Economic Research Service).

US Department of Agriculture, Economic Research Service, Office of Economics (1991) *Economic Implications of the Uruguay Round for U.S. Agriculture* (Washington, DC: US Department of Agriculture, May).

US General Accounting Office (1992) *Progress in GATT Negotiations* (Washington, DC: General Accounting Office).

US General Accounting Office (1994) *International Trade: Impact of the Uruguay Round Agreement on the Export Enhancement Program, Briefing Report to the Honorable Thomas A. Daschle, U.S. Senate* (Washington, DC: General Accounting Office).

US International Trade Commission (1990a) *Estimated Tariff Equivalents of Non-tariff Barriers on Certain Agricultural Imports in the European Community, Japan, and Canada: Supplemental Report to the President on Investigation No. 332–281 under Section 332(g) of the Tariff Act of 1930, as amended* (Washington, DC: USITC Publication 2280, April).

US International Trade Commission. (1990b) *Estimated Tariff Equivalents of U.S. Quotas on Agricultural Imports and Analysis of Competitive Conditions in U.S. and Foreign Markets for Sugar, Meat, Peanuts, Cotton, and Dairy Products: Report to the President on Investigation No. 332–281 under Section 332(g) of the Tariff Act of 1930, as amended* (Washington, DC: USITC Publication 2276, Part II of II, April).

US Statutes at Large 49 (1936) Agricultural Adjustment Act, amendments, p. 774.

US Statutes at Large 102 (1988) Omnibus Trade and Competitiveness Act of 1988.

US Statutes at Large 104 (1990) Omnibus Budget Reconciliation Act of 1990.

US Statutes at Large 107 (1993) Extension of Uruguay Round Trade Agreement Negotiating and Proclamation Authority and of 'Fast Track' Procedures to Implementing Legislation, pp. 239–40.

US Trade Representative (1984–94) Speeches and Testimonies File (Washington, DC: Office of the United States Trade Representative).

US Trade Representative (1994) *Uruguay Round of Multilateral Trade Negotiations, General Agreement on Tariffs and Trade* (Washington, DC: Office of the United States Trade Representative).

'United States Proposal of 6 July 1987 for Negotiations on Agriculture' (1988) in Petersmann and Hilf (eds), pp. 585–8.

Uriu, Robert M. (1996) *Troubled Industries: Confronting Economic Change in Japan* (Ithaca, NY: Cornell University Press).

Vahl, Remco (1997) *Leadership in Disguise: The Role of the European Commission in EC Decision-Making on Agriculture in the Uruguay Round* (Brookfield, Vt.: Ashgate).

Verdier, Daniel (1994) *Democracy and International Trade: Britain, France, and the United States, 1860–1990* (Princeton, NJ: Princeton University Press).

VerLoren van Themaat, Pieter (1981) *The Changing Structure of International Economic Law* (Boston: Mass.: Martinus Nijhoff).

Vernon, Raymond and Debora Spar (1989) *Beyond Globalism: Remaking American Foreign Economic Policy* (New York: Free Press).

Vernon, Raymond, Debora L. Spar and Glenn Tobin (1991) *Iron Triangles and Revolving Doors: Cases in U.S. Foreign Economic Policymaking* (New York: Praeger).

Vogel, Steven K. (ed.) (2002) *U.S.–Japan Relations in a Changing World* (Washington, DC: Brookings Institution).

Wallerstein, Immanuel M. (1974) *The Modern World-System* (New York: Academic Press).

Wallerstein, Immanuel M. (1989) *The Modern World-System III: The Second Era of Great Expansion of the Capitalist World-Economy, 1730–1840s* (New York: Academic Press).

Waltz, Kenneth (1967) *Foreign Policy and Democratic Politics* (Boston, Mass.: Little, Brown).

Waltz, Kenneth (1979) *Theory of International Politics* (New York: McGraw-Hill).

Ward, Robert E. and Yoshikazu Sakamoto (1987) *Democratizing Japan: The Allied Occupation* (Honolulu, HI: University of Hawaii Press).

Watts, William (1984) *The United States and Japan: A Troubled Partnership* (Cambridge, Mass.: Ballinger Publishing).

Weatherford, M. Stephen (1988) 'The International Economy as a Constraint on U.S. Macroeconomic Policymaking', *International Organization*, vol. 42 (Autumn), pp. 605–37.

Winham, Gilbert R. (1986) *International Trade and the Tokyo Round Negotiation* (Princeton, NJ: Princeton University Press).

Winham, Gilbert R. and Ikuo Kabashima (1982) 'The Politics of U.S.–Japanese Auto Trade', in Destler and Sato (eds), *Coping with U.S.–Japanese Economic Conflicts*, pp. 73–119.

Wittkopf, Eugene and James M. McCormick (eds) (1999) *The Domestic Sources of American Foreign Policy, Insights and Evidence*, 3rd edn (Lanham, Md.: Rowman & Littlefield).

Wolfe, Robert (1998) *Farm Wars: The Political Economy of Agriculture and the International Trade* Regime (New York: St. Martin's Press).

World Bank (1993) *The East Asian Miracle: Economic Growth and Public Policy: A World Bank Policy Research Report* (New York: Oxford University Press).

World Trade Organization (1997) *WTO Trade Policy Review Series, United States, 1996* (Geneva: World Trade Organization).

World Trade Organization (1998) *WTO Trade Policy Review Series, Japan, 1998* (Lanham, Md.: Bernan).

'Written Responses to Questions Submitted by the Japanese Newspaper *Asahi Shimbun*, 28 April 1987', in Reagan, *1987, Book I – January 1 to July 3, 1987*, 1989, pp. 429–33.

Yabunaka, Mitoji (1991) *Taibei Keizai Kōshō: Masatsu no Jitsuzō, In Search of New Japan–U.S. Economic Relations, Views from the Negotiating Table* (Tokyo: Simul Press).

Yamaji, Susumu and Shoichi Ito (1993) 'The Political Economy of Rice in Japan', in Tweeten, Dishon, Chern, Imamura and Morishima (eds), pp. 349–65.

Yamamura, Kozo (ed.) (1982) *Policy and Trade Issues of the Japanese Economy: American and Japanese Perspectives* (Seattle, Wash.: University of Washington Press).

Yamamura, Kozo (1994) 'The Deliberate Emergence of a Free Trader: The Japanese Political Economy in Transition', in Craig Garby and Mary Brown Bullock (eds), *Japan: A New Kind of Superpower?* (Baltimore, Md.: Johns Hopkins University Press).

Yamamura, Kozo (1995) 'The Role of Government in Japan's "Catch-Up" Industrialization: A Neoinstitutionalist Perspective', in Hyung-Ki Kim, Michio Muramatsu, T. J. Pempel and Kozo Yamamura (eds), *The Japanese Civil Service and Economic Development: Catalysts of Change* (New York: Oxford University Press), pp. 102–32.

Yamamura, Kozo (ed.) (1997) *The Economic Emergence of Modern Japan* (New York: Cambridge University Press).

Yamamura, Kozo and Yasukichi Yasuba (eds) (1987) *The Political Economy of Japan, Vol. 1: The Domestic Transformation* (Stanford, Calif.: Stanford University Press).

Yamazawa, Ippei (1992) 'Gearing Economic Policy to International Harmony', in Hook and Weiner (eds), *The Internationalization of Japan*, pp. 119–30.

Yarbrough, Beth and Robert M. Yarbrough (1987) 'Cooperation in the Liberalization of International Trade', *International Organization*, vol. 41 (Winter), pp. 1–26.

Yarbrough, Beth and Robert M. Yarbrough (1992) *Cooperation and Governance in International Trade: The Strategic Organizational Approach* (Princeton, NJ: Princeton University Press).

Yoffie, David B. (1986) 'Protecting World Markets', in McCraw (ed.), *America versus Japan*, pp. 35–75.

Yoffie, David B. (1988) 'American Trade Policy: An Obsolete Bargain?', in Chubb and Peterson (eds), *Can the Government Govern?* (Washington, DC: Brookings Institution), pp. 100–38.

Young, Michael K. and Yuji Iwasawa (eds) (1996) *Trilateral Perspectives on International Legal Issues: Relevance of Domestic Law and Policy* (Irvington, NY: Transnational Publishers).

Young, Oran R. (1994) *International Governance: Protecting the Environment in a Stateless Society* (Ithaca, NY: Cornell University Press).

Young, Oran R. (1998) *Creating Regimes: Arctic Accords and International Governance* (Ithaca, NY: Cornell University Press).

Zartman, I. William and Maureen R. Berman (1982) *The Practical Negotiator* (New Haven, Conn.: Yale University Press).

Zartman, I. William (ed.) (1994a) *International Multilateral Negotiation: Approaches to the Management of Complexity* (San Francisco, Calif.: Jossey-Bass).

Zartman, I. William (1994b) 'Introduction: Two's Company and More's a Crowd: The Complexities of Multilateral Negotiation', in Zartman (ed.), *International Multilateral Negotiation: Approaches to the Management of Complexity*, pp. 1–10.

Zartman, I. William (1997) 'Negotiating Regime Evolution in the Environmental Field: The Dynamics of Regime Building' (Paper presented at the annual meeting of the American Political Science Association, Washington, DC, August).

Zartman, I. William and Jeffrey Z. Rubin (eds) (2000) *Power and Negotiation* (Ann Arbor, Mich.: University of Michigan Press).

Zenkoku Nōgyō Kyōdō Kumiai Chūōkai (Central Union of Agricultural Co-operatives) (1988) *Gatto to Nōgyō* (*GATT and Agriculture*) (Tokyo: Tsukuba Shobō).

Zysman, John (1983) *Governments, Markets and Growth: Financial Systems and the Politics of Industrial Change* (Ithaca, NY: Cornell University Press).

Zysman, John and Laura Tyson (eds) (1983) *American Industry in International Competition: Government Policies and Corporate Strategies* (Ithaca, NY: Cornell University Press).

Index

Aaronson, Susan, 22, 149
Adams, L. Jerold, 28
agricultural export subsidies, 45, 49, 85, 101, 129, 139
agricultural interest groups
  in Japan, 33, 71, 72, 94
  in US, 21, 48, 64, 66–8, 138–40
  worldwide, 115–16
agricultural market access, 59, 85, 113, 124, 126, 141
  see also market access
agricultural reform, 66, 101, 104, 140
  in Japan, 69, 79–80, 90–1, 94–6, 143
agricultural subsidies, 44, 48, 65, 106, 111, 118
Akaneya, Tatsuo, 30
Allinson, Gary, 149
American Agriculture Movement, 57
American Corn Growers Association, 62, 64
American Farm Bureau Federation, 48, 50, 56, 57, 59, 67, 141
American Sugar Alliance, 55
Amstutz, Daniel, 109
  USDA Under Secretary for International Affairs and Commodity Programs, 45, 164
Arase, David, ix
Argentina, 99, 102, 107, 118
Asahi Shimbun (newspaper), 105, 125, 156–75
ASEAN member countries, 115
Associated Milk Producers, 52
Association of Corporate Executives (*Dōyūkai*), in Japan, 83
Australia, 55, 99, 101, 102, 105, 107, 118
Austria, 115, 116, 118, 119
Avery, William, viii, 3, 175

Baldwin, Robert, 9, 16, 22, 138, 139
Bangladesh, 116
barley, 43, 81, 125

basic foodstuffs, 79, 107, 109, 114, 118, 126
Baucus, Max, US Senator, 53, 139
beef, 38, 57, 71, 76, 78, 82, 84, 98
Bhala, Raj, 19
Blaker, Michael, 31, 129
Bradley, Bill, US Senator, 100
Brazil, 107, 116
Breen, John, viii, 104, 107, 122, 126
Britain, 21
Brittan, Sir Leon, EC External Economic Affairs Commissioner, 126
Broadbent, Jeffrey, 149
Brock, William, USTR, 43, 44
Bush, George H. W.
  Administration of (January 1989–January 1993), ix, 50–60, 66
  US President, 55

Cairns Group, 58, 62, 101, 103, 104, 106–7, 108, 114–18, 130
Calder, Kent, 31
Cameron, Charles, ix
Canada, 20, 61, 99, 101–2, 104, 106–7, 115, 118–20, 123–4
capitalism, 8, 10, 11, 20, 30, 145
Caporaso, James, 1–2, 148
case studies, 5, 39–40, 135, 139, 142, 145, 147
Central Union of Agricultural Co-operatives (*Zenchū*), in Japan, 72, 74, 77, 81, 83, 87, 96, 115, 132, 141
Chalmers, Douglas, viii
Chiba, Kazuo, 164, 167
Chile, 99, 107
Christensen, Ray, 34
citrus, 38, 43, 70–1, 75–6, 78, 82
Clinton, William
  Administration of (January 1993–January 2001), 4, 23, 60–5, 66, 67, 123
  US President, 63–4
Coalition to Save the Family Farm, 50

Janow, Merit, 38
Japan, vii, viii, ix, 3–5, 11, 14–15,
    25–38, 40, 43, 46, 48, 49, 59, 60,
    61, 64–5, 66, 68, 69–97, 98–133,
    135–6, 140, 141–7
Japan, Agricultural Policy Council, 73,
    90
Japan, Cabinet of, 28–9, 86, 89, 135
Japan, Ministry of Agriculture,
    Forestry and Fisheries, 27, 72–4,
    77, 81, 84, 88, 90–2, 94–6, 112,
    114, 122, 131, 141–3
    see also Eijirō Hata; Mutsuki Katō;
    Masami Tanabu; Tomio
    Yamamoto
Japan, Ministry of Economy, Trade
    and Industry, 26–8
Japan, Ministry of Education
    (*Monbushō*), viii–ix, 28
Japan, Ministry of Finance, 27, 72, 77,
    90–1, 94, 102, 142–3
Japan, Ministry of Foreign Affairs, 27,
    73, 77, 90–2, 95, 102, 131, 141, 143
    see also Kuranari; Nakayama; Uno;
    Watanabe
Japan, Ministry of International Trade
    and Industry, 26–7, 32, 77, 90–2,
    94–5, 102, 131, 141–3
    see also Hajime Tamura
Japan Agrinfo Newsletter, 157, 158, 161,
    166, 167, 170
Japan Communist Party, 71, 76, 79,
    83–4, 88–9, 96
Japan New Party, 85–6, 91
Japan Renewal Party (*Shinseitō*), 85–6,
    88–9, 93
Japan Socialist Party (JSP), 71, 76,
    78–9, 83–4, 86–7, 89, 93, 96, 131
Japan Times, 159, 166–8
Japan's food control system, 72, 79,
    90–2, 94, 132
Japan's negotiating stance in the
    Uruguay Round agriculture
    negotiations, 74–97, 104, 109,
    125, 128–33, 141–4, 146–7
Japan's rice market, 49, 60, 72, 77–8,
    80–2, 86–7, 90–2, 105, 122, 125, 130
Japan's self-sufficiency, 31, 70–1, 76–7,
    81, 84, 131

Japan's virtual ban on rice imports,
    57, 61, 70–1, 75, 80, 86, 91, 94,
    102, 105, 111, 114, 116, 120, 122,
    125, 129, 131, 142, 144
Japanese agricultural policy, 72–3, 76,
    78, 90
Japanese beef and citrus, 71, 76, 82
Japanese bureaucracy, 26–8, 32, 35,
    90–1, 94, 131, 141–2, 144, 146
Japanese Constitution, 28, 32
Japanese Diet, 26, 28, 32–3, 35, 39, 70,
    73, 76–8, 85–6, 89, 91–4, 97, 128,
    131–3, 141–3, 146
    resolutions of opposing liberalization
    of Japan's rice market, 71, 76,
    80, 87, 92, 93, 133
Japanese Diet, Lower House, 70, 72,
    82
    approval of Uruguay Round
    agreements, 89
    February 1990 election, 79, 84
    July 1986 election, 72
    July 1993 election, 85, 96
    rice self-sufficiency resolution of
    1984, 71
Japanese Diet, members of, 28, 90–6,
    131, 135, 141, 144
Japanese Diet, Upper House
    July 1989 election, 78, 84, 94–5
    July 1992 election, 84, 96
    LDP majority in, 78, 96
Japanese farmers, 70–1, 75–6, 78–9,
    83, 91, 96, 130, 132, 142–3,
Japanese government, 28, 32, 35, 39,
    46, 70–2, 75–9, 81, 84, 86–7,
    89–93, 119
Japanese Prime Minister, 26–8, 92
    see also Hata; Hosokawa; Kaifu;
    Miyazawa; Murayama;
    Nakasone; Takeshita; Uno
Japanese rice, 58, 69–74, 77–80, 81,
    83–4, 88, 109, 125
Japanese rice market liberalization, 34,
    75–97, 104–5, 108–9, 121,
    129–33, 142–4
Japanese trade policymaking, vii, 14,
    25–35, 69–87, 141–4, 147, 149
Johnson, Chalmers, 25, 27, 32, 141
Johnson, D. Gale, 43, 100

Domestic Politics and International Relations
in US–Japan Trade Policymaking

International Political Economy Series

General Editor: **Timothy M. Shaw**, Professor of Commonwealth Governance and Development, and Director of the Institute of Commonwealth Studies, School of Advanced Study, University of London

Titles include:

Lucian M. Ashworth and David Long (*editors*)
NEW PERSPECTIVES ON INTERNATIONAL FUNCTIONALISM

Robert W. Cox (*editor*)
THE NEW REALISM
Perspectives on Multilateralism and World Order

Frederick Deyo (*editor*)
GLOBAL CAPITAL, LOCAL LABOUR

Stephen Gill (*editor*)
GLOBALIZATION, DEMOCRATIZATION AND MULTILATERALISM

Björn Hettne, András Inotai and Osvaldo Sunkel (*editors*)
GLOBALISM AND THE NEW REGIONALISM

Christopher C. Meyerson
DOMESTIC POLITICS AND INTERNATIONAL RELATIONS IN US–JAPAN TRADE POLICYMAKING
The GATT Uruguay Round Agriculture Negotiations

Michael G. Schechter (*editor*)
FUTURE MULTILATERALISM
The Political and Social Framework

Michael G. Schechter (*editor*)
INNOVATION IN MULTILATERALISM

Thomas G. Weiss (*editor*)
BEYOND UN SUBCONTRACTING
Task Sharing with Regional Security Arrangements and Service-Providing NGOs

Robert Wolfe
FARM WARS
The Political Economy of Agriculture and the International Trade Regime

International Political Economy Series
Series Standing Order ISBN 0–333–71708–2 hardback
Series Standing Order ISBN 0–333–71110–6 paperback
(*outside North America only*)

You can receive future titles in this series as they are published by placing a standing order. Please contact your bookseller or, in case of difficulty, write to us at the address below with your name and address, the title of the series and one of the ISBNs quoted above.

Customer Services Department, Macmillan Distribution Ltd, Houndmills, Basingstoke, Hampshire RG21 6XS, England